PRAIS
The Dance o

"Harriet Lerner has outdone herself, jumping into an area we fear most—speaking truthfully to loved ones who hurt us. I treasure this book and used it immediately—and so will you."

—Betty Carter, M.S.W., author of *Love, Honor and Negotiate*

"Profound empathy. . . . Lucid and concrete guidance."

—*Publishers Weekly*

"I love Harriet Lerner's work."

—Anne Lamott, author of *Traveling Mercies*

"*The Dance of Connection* can save your marriage, a friendship, and your relationship with your mother, father, sister, brother— even your boss. Lerner is one of the wisest, most sophisticated psychology writers alive today. Her book is brimming with practical advice, sharp wit, extraordinary knowledge, deep caring, and a wealth of ah-yes-how-true wisdom into human nature. Lerner knows us all so well."

—Edward M. Hallowell, M.D., author of *Driven to Distraction*

"Accessible and well organized. . . . Highly recommended."

—*Library Journal Review*

© Chris Ochsner

About the Author

HARRIET LERNER, PH.D., is one of our nation's most loved and respected relationship experts. Renowned for her work on the psychology of women and family relationships, she served as a staff psychologist at the Menninger Clinic for more than two decades. A distinguished lecturer, workshop leader, and psychotherapist, she is the author of *The Dance of Anger* and many other bestselling books. She is also, with her sister, an award-winning children's book writer. She and her husband live in Kansas and have two sons.

BY THE AUTHOR

The Dance of Connection

How to Talk to Someone When
You're Mad, Hurt, Scared, Frustrated,
Insulted, Betrayed, or Desperate

Harriet Lerner, Ph.D.

HARPER

NEW YORK · LONDON · TORONTO · SYDNEY

A hardcover edition of this book was published in 2001 by HarperCollins Publishers.

THE DANCE OF CONNECTION. Copyright © 2001 by Harriet Lerner. All rights reserved. Printed in the United States of America. No part of this book may be used or reproduced in any manner whatsoever without written permission except in the case of brief quotations embodied in critical articles and reviews. For information address HarperCollins Publishers, 195 Broadway, New York, NY 10007.

HarperCollins books may be purchased for educational, business, or sales promotional use. For information, please e-mail the Special Markets Department at SPsales@harper-collins.com.

First Quill edition published 2002.

Designed by The Book Design Group

The Library of Congress has catalogued the hardcover edition as follows:

Lerner, Harriet.
The dance of connection : how to talk to someone when you're mad, hurt, scared, frustrated, insulted, betrayed, or desperate/Harriet Lerner.
    p.   cm.
  ISBN 0-06-019638-6
  Includes bibliographical references.
  1. Interpersonal relations. 2. Intimacy (Psychology). I. Title.

BF637.C45 L47 2001
153.6—dc21   00-054182

ISBN 0-06-095616-X (pbk.)

23 24 25 26 27 LBC 43 42 41 40 39

To my best friends

Contents

ACKNOWLEDGMENTS

My close friends cheered me on, read and reread this entire manuscript with love, intelligence, and care, and offered sharp criticism and expert advice. My heartfelt thanks to Jeffrey Ann Goudie, Emily Kofron, Marcia Cebulska, Marianne Ault-Riché, Stephanie von Hirschberg (who also supplied the title *The Dance of Connection*), and Tom Averill. I'm especially indebted to Mary Ann Clifft at Menninger, a meticulous and generous editor who reviewed every page and compiled the index.

My love and gratitude go to my friend/agent/manager extraordinaire, Jo-Lynne Worley, for her remarkable competence, unwavering faith in my work, and day-to-day support since she teamed up with me in 1990. It has also been my good fortune to be wedded to HarperCollins, the publisher of all my books to date and the home of the best team of people with whom I could ever hope to work. My editor, Gail Winston, is an author's dream; she has kept me on track by offering the perfect balance of criticism and regard.

Thanks to my husband, Steve, for everything—and most of all for loving me so much and for making me laugh so hard even when I'm in the grumpiest of moods. The fact that he came into my life with a cadre of in-laws, out-laws, and extended family who exhaust my vocabulary of superlatives is quite a perk.

Other people have critiqued parts of this manuscript or surrounded me with their wisdom, affection, or practical help during the time I was writing it. Thanks to my niece/editor, Jen Hofer; my sister, Susan Goldhor; my mother, Rose Goldhor; and my two sons, Matt and Ben, who are in my heart every day. Thanks to my publicist Jessica Jonap and also to Vonda Lohness, Ellen Safier, Nancy Maxwell, Julie Cisz, Marilyn Mason, Miranda Ottewell, Libby Rosen, and especially to Joanie Shoemaker. Professional acknowledgments appear at the end of the book.

Over the years, women and men in psychotherapy have shared their stories with me, readers have overwhelmed me with expressions of gratitude and affection, and friends and colleagues too numerous to be named have nourished, supported, informed, and sustained me. All these people have made my life larger and my writing possible.

Back to the Sandbox

I recently heard a story. Two little kids are playing together in a sandbox in the park with their pails and shovels. Suddenly a huge fight breaks out, and one of them runs away, screaming, "I hate you! I hate you!" In no time at all they're back in the sandbox, playing together as if nothing has happened.

Two adults observe the interaction from a nearby bench. "Did you see that?" one comments in admiration. "How do children do that? They were enemies five minutes ago."

"It's simple," the other replies. "They choose happiness over righteousness."

Grown-ups rarely make such a choice. We have a terrible time stepping aside from our anger, bitterness, and hurt. We know that life is short, but damn it, we're not getting back in that sandbox until that other person agrees to having started it—and admits to being wrong. Our need to balance the scales of justice is so strong that we lock ourselves into negativity at the expense of happiness and well-being.

A great deal of suffering could be avoided if we could be more like those kids. We could lighten up and let things go. I feel calmed and relieved when my husband knocks at my study door in the middle of a fight, puts his arms around me, and says, "I love you. This is stupid. Let's just drop it." Like two kids in the sandbox, we're suddenly light and playful again.

Of course, adult life is not always so simple. Some issues need to be revisited—not dropped—and talk is essential to this process. We need words to begin to heal betrayals, inequalities, and ruptured connections.

Our need for language, conversation, and definition goes beyond the wish to put things right. Through words we come to know the other person—and to be known. This *knowing* is at the heart of our deepest longings for intimacy and connection with others. How relationships unfold with the most important people in our lives depends on courage and clarity in finding voice. This is equally true for our relationship with our self. Even when we are not being heard, we may still need to know the sound of our own voice saying out loud what we really think.

Our challenge as adults is to develop a strong voice that is uniquely our own, a voice that reflects our deepest values and convictions. Once we are comfortable within that voice, we can bring it to our most important relationships. We can choose to move to the center of a difficult conversation—or we can let it go. We can speak—or decide not to. Whatever we choose, we can head back to the sandbox with clarity, wisdom, and intention. By doing so, we can strengthen the self and our connections, and have the best chance of achieving happiness during our time with each other.

CHAPTER 1

Finding Your Voice

The thread that unites my work both as an author and as a psychotherapist is my desire to help people speak wisely and well, sometimes about the most difficult subjects. This includes asking questions, getting a point across, clarifying desires, beliefs, values, and limits. How such communication goes determines whether we want to come home or stay away at the end of the day.

This is no simple matter, as glib terms like *communication skills* or *assertiveness training* imply. Assertiveness is considered a good idea—if not a cultural ideal. But despite decades of assertiveness training and lots of good advice about communicating with clarity, timing, and tact, we may do our best to speak but still feel unheard. We may find that we cannot affect our husband or wife or partner, that fights go nowhere, that conflict brings only pain rather than an opportunity for two people to learn more about each other. We may have the same dilemma with our mother, sister or uncle, or close friend.

The Limits of "Good Communication"

We all want to communicate well and make ourselves heard. "He just doesn't get it" or "She's so critical" are sentiments I hear daily in my work. When we speak from the heart, we long for an ear to hear us, and we all have experienced that down feeling when we perceive ourselves as written off or misunderstood.

I wish I could reassure you that reading this book will guarantee that you will finally be heard in your most difficult relationships. Or that strengthening your voice will bring you the love and approval of others. Or that following my good advice will give you a deep sense of inner peace.

Truth is, nothing you say can ensure that the other person will get it, or respond the way you want. You may never exceed his threshold of deafness. She may never love you, not now or ever. And if you are courageous in initiating, extending, or deepening a difficult conversation, you may feel even more anxious and uncomfortable, at least in the short run.

All the assertiveness training and communication skills in the world can't prevent a relationship from becoming fertile ground for silence and stonewalling, or for anger and frustration, or for just plain hard times. No book or expert can protect us from the range of painful emotions that make us human. We can influence the other person through our words and silence, but we can never control the outcome.

That said, what we *can* learn in the chapters ahead is enormous. We can *maximize* the chance of being heard and moving relationships forward. We can take a conversation to the next level when the initial foray doesn't bring the desired result. We can stop nonproductive conversational habits so that an old relationship will take a new turn. We can clarify what we feel entitled to and responsible for—and what we really want to say. Or, alternatively, we can learn to sit more comfortably with our confusion. We can operate from a solid position of self, even when the other person won't speak to us at all.

TOWARD AN AUTHENTIC VOICE

The challenge of finding an authentic voice within an intimate relationship is far larger than a word like *communication* can ever begin to convey. Authenticity brings to mind such elusive qualities as being *fully present, centered,* and *in touch* with our best selves in our most important conversations. Moving in this direction requires us to clarify—to ourselves and others—what's important to us. Having an authentic voice means that:

- We can openly share competence as well as problems and vulnerability.

- We can warm things up and calm them down.

- We can listen and ask questions that allow us to truly know the other person and to gather information about anything that may affect us.

- We can say what we think and feel, state differences, and allow the other person to do the same.

- We can define our values, convictions, principles, and priorities, and do our best to act in accordance with them.

- We can define what we feel entitled to in a relationship, and we can clarify the limits of what we will tolerate or accept in another's behavior.

- We can leave (meaning that we can financially and emotionally support ourselves), if necessary.

The second half of this list is about knowing our bottom line—that is, the values, beliefs, and priorities that are so crucial to preserving and protecting the self that we will not compromise them in any relationship. This is, perhaps, the most difficult challenge in couples.

In the abstract, any or all of these actions may seem obvious and easy. But when we are dealing with difficult subjects or significant relationships, nothing is ever simple.

BOLD NEW CONVERSATIONS

The challenge in conversation is not just to be our self but to choose the self we want to be. What we call "the self" is never static, but instead is a work in progress. That's why we don't discover who we are by sitting alone on a mountaintop and meditating, or by being introspective and "going deeper," as valuable as these disciplines may be. The royal road for both discovering and reinventing the self is through our relationships with other people and the conversations we engage in.

In a sad paradox, the more important and enduring a relationship (say, with a partner or relative), the more we tend to participate in narrow, habitual conversations where our experience of our self and the other person becomes fixed and small. My goal is to challenge us to engage in novel conversations that will create a larger, more empowering view of who we are and what is truly possible.

Although I resonate with the phrase "finding our voice," the image it evokes is deceptive. We don't dig our authentic voice out of the muck, as a dog digs through the dirt to uncover a prized bone. Similarly we don't just reveal ourselves in conversation; we can also discover and deepen who we are. The challenge then, is not only to find our authentic voice but also to enlarge it.

THE PARADOX OF AUTHENTICITY

Speaking our mind and heart is the most precious of human rights. The ability to speak our own truths forms the core of both intimacy and self-regard. The poet Adrienne Rich puts it beautifully: It is not, she writes, that we have to tell everything, or to tell it all at once, or even to know beforehand all that we need to tell. But an honorable relationship, she reminds us, is one in which "we are trying, all the time, to extend the possibilities of truth between us . . . of life between us." When we are not able to speak authentically, our relationships spiral downward, as does our sense of integrity and self-regard.

We all long to have a relationship so relaxed and intimate that we can share anything and everything without first thinking about it. Who wants to hide out in a relationship in which we can't allow ourselves to be known? Speaking in our own voice, not in someone else's, is an undeniably good idea. I've yet to meet the person who aspires to be phony or invisible in her closest relationships. The dictate "Be yourself " is a cultural ideal touted everywhere, and luckily, no one else is as qualified for the job.

But therein lies the paradox

Speaking out and being "real" are not necessarily virtues. Sometimes voicing our thoughts and feelings shuts down the lines of communication, diminishes or shames another person, or makes it less likely that two people can hear each other or even stay in the same room. Nor is talking always a solution. We know from personal experience that our best intentions to process a difficult issue can move a situation from bad to worse. We can also talk a particular subject to death, or focus on the negative in a way that draws us deeper into it, when we'd be better off distracting ourselves and going bowling.

A marital therapist recently teased me, "Are you writing *another* book to help women speak up? I'm trying to help my clients be quiet." Then she said more seriously, "Why do people think they have to tell each other *everything* they feel? Why must they share their uncensored reactions?" She was referring to the corrosive criticism that wears couples down as they selectively attend to what bothers them in a partner rather than speaking to what they appreciate and admire. And she was referring to the raw, unbridled emotional exchanges that, when unchecked, erode intimacy and connection in family relationships.

Indeed, in many situations wisdom lies in being strategic rather than spontaneous. This is especially true when we're dealing with a difficult person, a hot issue, or a tense situation. The enormous challenge is to make wise decisions about how and when to say what to whom, and even before that, to know what we really want to say and what we hope to accomplish by saying it.

In the chapters ahead we will be learning to edit, think, plan, and even pretend in our conversations—not out of fear and a desire for safety but rather from a position of courage and adventure. The goal is not to take flight from authenticity but to make a creative move toward it. Truth telling, like peacemaking, doesn't just bloom in our midst. Sometimes it has to be plotted and planned, a challenge that runs counter to the powerful urge to simply be our true, wonderful, spontaneous, uncensored selves.

DIFFICULT CONVERSATIONS

Obviously we're most challenged in difficult conversations. We feel angry, frustrated, confused, scared, hurt, insulted, betrayed, exhausted, or desperate. A relationship may reach a crisis point when we can't make ourselves heard, or (on the other side of the conversational fence) when we can't keep listening to another person who is wearing us down with complaints, criticism, negativity, demands, excuses, or irresponsible behavior. We may need an apology or acknowledgment that isn't forthcoming. As we learn to better navigate these conversations and silences better, we create opportunities to deepen our understanding of the self, the other, and our relationships. And we can learn when it's time to give up and get out—that is, to save ourselves—even if we end up losing a relationship.

This is a major challenge, or more accurately, a lifelong venture. We all bring varied levels of clarity and maturity to our conversations with different people, or even to the same relationship on different days. The challenge of having a voice is especially difficult and complex in the intense emotional field of family life. My younger son, Ben, once summed up this dilemma nicely when my husband, Steve, was driving him home from the airport during the winter break of his sophomore year of college. "You know, Dad," Ben said matter-of-factly, "as we get closer and closer to home, I can just feel the layers of maturity peeling off me."

People are my stock-in-trade, but when I'm anxious or angry enough, I have the brain of a reptile. For many adults, love and marriage may be the arena where maturity peels away or a fog descends on our brain and dissolves our thinking center. Or it may still be that a visit home makes us clutch so we can't creatively steer a conversation in a new direction.

Some individuals or groups enhance and empower us, while others do the opposite, so we may feel brilliant within one relationship but unable to articulate a clear thought in another. Many of the stories we take to be true or fixed about ourselves can change dramatically when we have conversations with people who make our world larger, not smaller. By doing our part to develop rather than diminish our voice, we can:

- **Create a more accurate and complex picture of ourselves and another person.**

- **Speak with honor and personal integrity even when the other person behaves badly.**

- **Strengthen our capacity for creativity, wisdom, joy, and zest.**

- **Increase our capacity to give and receive love.**

In sum, how we use our voice is at the heart of who we are in the world and the foundation of both intimacy and self-regard.

NOT JUST A WOMAN'S ISSUE

The Dance of Connection continues my long tradition of writing for women. Yet I certainly hope that men will read this book and find themselves well represented and richly rewarded. When it comes to the challenge of speaking wisely and well, we're all in this soup together.

I want to say up front that I don't agree with the popular notion that men, in general, are better than women at speaking up in a real way. True, it's the voices of privileged men that "really count," at least in the public sphere. Despite three decades of feminism, the world's major institutions are still run almost exclusively by men (although probably not by the particular guys we hang out with). But in their private lives, even those men often fall silent—or speak too loudly—when they feel they can't hold their own in talking things through.

Many men get caught between the cultural expectation that to "be a man" they must be dominant and in command, and their actual experience, which is one of voicelessness. Some men act out or act up in a variety of ways when they can't make themselves heard by the key people in their lives—a process that slowly erodes and eventually destroys their self-respect.

Getting beyond Stereotypes

Myth also has it that women need conversation and togetherness more than men do. We are told, for example, "Women seek connection, men seek separateness," or "Men want more sex, women want more talking." Such statements have a ring of truth, but a yin-yang psychology of opposites doesn't do justice to the complexity of human experience nor to the reality of any particular individual on any given day.

In marriage, for example, the man more frequently falls apart when a relationship ends, especially when his wife leaves him out of the blue. Often he has not really heard the seriousness of her complaints, paid attention to the entrenched distance between them, or allowed himself to be empathically affected by her telling him over and over that his actions (or inaction) cause her pain. That is, he hasn't been paying attention. As one man explained, "I've been going along stupid."

But the woman may also contribute to her partner's failure to see the impending breakup or divorce. She may complain, perhaps endlessly, but then continue to tolerate the status quo, going along with business as usual. At the point where the marriage or relationship is truly at risk, she may not do what it takes to get past his threshold of deafness, to explain how seriously she is thinking of leaving if he doesn't change. For far too long, she has said one thing with her words ("I can't live with this") and another with her behavior (she continues to live with it). Her voice lands in the middle of the floor, in the widening space between them, making it harder for him to catch on when she says, "I really mean it."

Women typically voice the most pain in marriage when they feel they can't influence their partner. These same women may then feel relieved to hear the experts say that no one can change another person who doesn't want to change. All too often this wisdom translates into an "anything goes" (or an "I can't do anything about it" attitude) that leads a relationship into a downward spiral. The notion that "men are from Mars"—the opposite rather than the neighboring sex—can lead a woman to lose her voice, make excuses for her partner, and tolerate behavior that is too costly to her own self. It also encourages the man to avoid the challenge of finding his own authentic voice and taking full responsibility for his choices.

When women equate requesting a behavioral change with trying to teach the proverbial pig to sing, we don't strengthen our voice. Instead, we get sucked in by the latest research findings about how male and female brains are different, so men can't really be expected to pick up their socks. It feels easier to give up and adapt to unfair circumstances, despite the enormous long-term toll of making such accommodations.

But voice problems also transcend gender, sexual orientation, and the other filters (race, class, generation) that color our view of the world. Anybody, for example, can encounter trouble being heard, or can fail to distinguish between complaining and establishing a deeply felt position for self—a position beyond compromise. Anybody can fail to take a conversation to the necessary next step. Anybody can speak up (or stay silent) at the expense of the other person or the self. Despite our tendency to polarize the sexes, on most days I believe that humans are more alike than different.

BRINGING IT ALL BACK HOME

In the house where I grew up, it was my father who didn't speak his mind. I've been thinking a great deal about this fact since his death in January 1998, shortly before his eighty-ninth birthday. Now that he's gone, I've been appreciating how profoundly his life has shaped my work, and how it has led me to the subject of this book. My father inspired me through his passion for language and words, and by being a witty and gifted speaker. But most of all, he influenced me by the fact that he couldn't (or didn't) speak out when it mattered most.

In an ironic twist of fate, my father actually lost his physical capacity to speak a year prior to his death. When we planned his memorial service in our home, my husband Steve suggested we play an audiotape of my father telling stories we had recorded a decade earlier, before he entered the world of nursing homes. At first I thought it an odd idea to have a dead person's voice kicking off his own memorial service—especially when that person had been locked in total silence during the final year of his life. But it proved to be a marvelous suggestion. The friends who gathered in our home that afternoon appreciated this opportunity to get to know my father, not just from our recollections, but also from hearing his deep, sonorous voice and his clever way with words. I was especially moved to hear the sound of my father's voice once more, and I knew that he would have applauded our bringing him onto center stage in this fashion.

Because this book has been stewing in my mind since my father's death, it seems fitting to share more about his life, at least as it relates to my subject. Of course, everyone's take on her parents (or anybody else's parents, for that matter) is necessarily fractured, subjective, and incomplete. But I do believe that what follows will put my passion for helping people find their voice in some perspective. It will also illustrate that this book is not merely about improving communication and connection. It's about avoiding the tragedy of losing one's self.

CHAPTER 2

Voice Lessons from My Father

Not all men want to run the show. I recognized this simple fact early in my childhood by observing my parents' marriage. My father, Archie, was the accommodating partner, while my mother, Rose, made all the decisions. She decided how money would be managed and spent, how my sister and I would be reared, what art would hang on the walls, and just about everything else. Archie allowed it to be her call whether he would have one egg or two, seconds on dessert, or whether he would get the burned piece of toast. I never heard him protest, not even when Rose packed him a left-over string-bean sandwich for lunch.

Whenever my sister, Susan, or I had a question about anything or needed permission to do something, we went to our mother. So did Archie, who seemed content with his childlike subservience. When I was grown, with a family of my own, my parents would visit us in Topeka, Kansas. "What would you like for breakfast, Daddy?" I'd inquire. "Ask Mother what I can have," he'd respond cordially, as if it were simply the natural order of things to pass the decision along to a higher authority.

My mother, for her part, claimed that she had no choice but to treat Archie like a child because he behaved like one. She believed that he needed to be watched because he had been spoiled by his mother and because, left to his own devices, he would make unwise decisions on every front. She was also convinced that he couldn't change and that not much could be expected of him. Archie was a full and active participant in maintaining this view of himself.

Yet no one who knew my mother would have described her personality as bossy, controlling, or domineering. She simply wasn't that type. She was modest, soft-spoken, and warm, a quiet and careful listener. Despite her towering competence, she preferred to be in the background, an appreciative audience to others. It was my father who loved to pontificate and hold court.

For this reason, it doesn't quite ring true—at least not in the literal sense—to describe my father as a man who had no voice. In fact, Archie had a deep, resonant voice, and he took pride in his distinguished, professorial manner of speech. A gifted, charming, and witty storyteller, he was the talker in social situations and at family gatherings. Occasionally, he "innocently" made outrageous statements that offended people and embarrassed my mother. "Look at the size of that woman's *behind!*" he would loudly exclaim, when the rear end in question was standing right in front of us. In this way, he both rebelled against my mother's control while reinforcing it, since his social gaffes only strengthened her belief that he required close monitoring.

In the private interior of marriage and family, Archie retreated into silence and classical music whenever talk moved beyond the superficial. Or he would disappear for entire evenings and weekends to his shop in the basement of our house in Brooklyn, where he built furniture and fixed things. My father loved company and conversation, yet he had no intimate relationships at all, no experience of truly knowing another person and being known. How this shaped the experience of those who shared his life is borne out by the following story.

My Father's Silence

The summer before my sophomore year in high school, I worked in upstate New York as the music and drama counselor in a camp for children with disabilities. I didn't get along with the camp director, but I loved the kids and the challenge of the work. It proved to be a great summer—that is, until the last day of camp, or more accurately until the last fifteen minutes, when I royally screwed things up.

Camp was over, and my father was coming to drive me home. I was waiting by my packed duffel bags when one of the guys from the grounds crew shouted, "Hey, Harriet, wanna drive the tractor?" I had never driven anything more complicated than a two-wheeler, so I don't know what possessed me to climb aboard, or what possessed him to climb down and shout directions from the ground and then disappear when I crashed into the camp director's car before locating the brake. The director was furious at me—okay, maybe he had a reason. He tore up my paycheck and told me I wouldn't get a penny for the entire summer's work. I had dedicated myself heart and soul to the campers, but it didn't occur to me to come to my own defense, or even to imagine I had one. I was clueless about such matters as car insurance, so I thought I had ruined this man's entire financial future and it was lucky for me that he wasn't sending me off to rot in prison. This was the scene my father entered—the smashed automobile, the outraged camp director, and me leaning against the tractor, shamed, humiliated, and tearful.

What I recall most vividly about the whole awful incident was that my father didn't say one word during the long drive home about what he had witnessed. I knew he was disapproving, but he didn't venture a question such as, "Harriet, what were you doing on that tractor?" or even "What happened?" My silence matched his own. I didn't say, "I'm sorry," since we weren't acknowledging that anything bad had actually happened that I might be sorry for. Nor did I offer a description or explanation of what had transpired. We never did talk about this incident, and I never got paid.

Communication went this way between us throughout my growing-up years in Brooklyn. When things were calm and superficial, he was charming and entertaining. Susan and I have wonderful memories of family dinners where we would delight in each other's company and conversation. But if the subject at hand was emotionally loaded, my father would shut down.

In the world of things, Archie was truly at home and remarkably gifted. As Susan said at his memorial service, "The kind of intuitive understanding that most of us have for friendship, for family relations, for collegial getting along, Archie had for motors, for power tools, for electricity, plumbing, and construction, for how all things were put together and how they could be taken apart." He could fix or build anything without any instruction and seemingly without trial and error—an ability that seemed magical to my sister and me. But relationships had him stumped when they moved beyond bantering.

"It's Not Worth the Fight"

More than anything, my father hated a fight. He simply refused to engage in conflict, and he'd leave the room if he saw it coming. I never once heard him raise his voice or directly express anger. My mother, who had grown up frightened of her own father's temper, was initially drawn to Archie's gentle manner, his dedication to keeping the peace at any price. But the qualities that initially draw two people together are oftentimes the very same ones that they later complain about. "He won't fight," my mother would say. "He won't even discuss it," the *it* being any subject where feelings might be involved.

It wasn't just anger and conflict that my father avoided. Remarkably, he claimed never to have experienced the whole range of unpleasant emotions, including anxiety, fear, sadness, depression, or simple old worry. Not surprisingly, he was at a total loss to recognize or respond to the emotional life of others. My father's assertions of voice, his expressions of protest or resistance, his acknowledgment that all was not right—these were acted out indirectly rather than spoken.

For example, when my parents were both forty-seven years old, my mother was diagnosed with a serious cancer about which they never spoke. By my mother's reports, Archie avoided the subject totally, never once initiating a conversation about her illness and refusing to be engaged. Back then the cultural context encouraged silence and denial, and the word *cancer* was rarely mentioned. But even considering the repressive atmosphere of the 1950s, my father's silence was extreme.

But during Rose's long treatment and recovery, Archie kept the cancer in focus with a pair of rubber gloves that he wore when he cleaned the pots after dinner. Rose associated these gloves with her illness and hospital experience, and she asked him to please put them away when he was done with the dishes. My father, who was extremely tidy and organized, put everything away by habit. But like clockwork, he'd leave the despised rubber gloves out on the counter to greet my mother. If I had been Rose, I would have tossed them in the trash, but they remained a hot spot between them. He responded to her daily confrontation about leaving the gloves out with exactly the same two words. "I forgot," he'd say. My mother, for her part, kept the process going by intensifying her negative focus on Archie and the gloves, and by complaining to Susan and me about him.

* * *

On matters large and small, Archie often "forgot" to do the right and responsible thing. Words such as *passive-aggressive* or *manipulative* fit his behavior, but these pejorative terms fail to do justice to what drives us. We won't need to be passive-aggressive if we feel empowered to express our anger or will directly. Nor will we resort to manipulation if our past experience has encouraged us to speak frankly. Archie could not voice or even recognize his fear of losing Rose. Nor could he acknowledge his associated rage and grief about the possibility.

As an adult, I asked my father to think back on that difficult time, sharing my experience as a frightened twelve-year-old and inquiring about his experience. But each time I raised the issue, he claimed to have no response at all. "Weren't you a bit worried?" I'd venture, reminding him of the doctor's pessimistic prognosis, which gave Rose something like a year to live. My mother was so central to the emotional and functional life of our family that it was difficult to imagine how we could have gone on without her. "No," my father said, "I never worried. I assumed she'd be fine." In this case, he was right. My mother, now ninety-two, has outlived him.

It wasn't just the hot issues that Archie avoided. More to the point, my father didn't express his wants and beliefs or say anything that would bring the differences between himself and another person into bold relief and potentially disrupt the harmony. I don't recall his ever saying, "No, Rose, I don't agree with that," and then holding firm to his belief. Instead, he'd say nothing, accommodate on the surface, and then secretly do as he pleased. When I'd ask him what stood in the way of his speaking his mind, or saying "no" to Rose, he'd say, "It's not worth the fight."

My father, who was so good with words, was also terrified of them. "Words once spoken can never be taken back," he would say to me when I encouraged him to express himself. "But, Daddy," I would protest, even as a kid, "of *course* words can be taken back." I never grasped the logic of this platitude. Not that we can swallow our words back down our throat like a cartoon running in reverse, but I knew that words could be taken back because I took them back all the time. I'd shoot off my mouth, and then later apologize and explain, and everything would be fine.

But my father never distinguished between nonproductive fighting and taking a firm position on his own behalf. He assumed a de-selfed position in all his key relationships, meaning that his wants, beliefs, priorities, and values became negotiable under relationship pressures. My father chose to have relationships at the expense of having a self, a pattern that began long before he met and married Rose.

The Larger Family Picture

My parents' marriage was a natural fit. My mother, the eldest child in her family of origin, was the leader in her sibling group, while my father was the follower in his, tagging along even as a teenager so closely to his strong-willed big sister that they were often mistaken for boyfriend and girlfriend. From this perspective alone, the ease with which Rose took over and Archie accommodated made a certain amount of sense. But birth order is just one thread in the rich tapestry of family life and by itself could not account for my father's extreme accommodation.

Archie and Rose were both children of Russian Jewish immigrant parents who faced immeasurable hardships, beginning with their traumatic emigration from the old country. Both were fiercely loyal to their mother and distant from their father. Both were their mother's "special child." But beyond these commonalities, each of my parents occupied a very different position in the emotional life of their first families.

Rose was "the responsible one," "the good one," and her mother's closest confidante. She was a natural caretaker who basically reared her three younger siblings after her own mother developed health problems as a young woman and later died of tuberculosis at forty-four. Taking charge came easily to my mother, who quietly and without complaint did everything that needed to be done. In her own words, she was "never a child."

These bare facts might suggest a life of overresponsibility and deprivation, but there's never a trace of martyrdom or self-pity in my mother's stories about her past. Rather, she speaks of her relatives with a love and warmth that can only leave one feeling proud to be a member of this remarkable clan, which included a large, colorful extended family. The more difficult the circumstances, the more this proud, close-knit immigrant family drew together. "Family is everything" was their credo.

In Archie's family, the anxious aftermath of the immigration was not managed through family togetherness. To the contrary, it was as if family members replayed the traumatic losses they had experienced by turning passive into active—that is, by replicating the pattern of loss in their family relationships in the new country. Anyone who got mad would disconnect from the relationship, never to return. Differences weren't tolerated, so if you fought with a family member, that person might never forgive you, speak to you, or even recognize your existence. Considering what my father observed in the first and most influential group he ever belonged to, it's no wonder his watchword became, "Words once spoken, can never be taken back."

Archie, the first son and the middle of three children, was also his mother's favorite. But he did not hold a family position of responsibility. By all reports, he was babied, shamelessly spoiled, and kept close to home. The intensity of my grandmother's involvement with him was no doubt fueled by the fact that she was cut off from every single member of her family of origin, and later from Archie's father and her only daughter. Archie and his mother were joined at the hip.

Archie was his mother's loyal son. It was his job to wave her banner, fill up her empty bucket, and be her staunch ally against his father in a bitter marriage that never went well and didn't last long. Archie has no recollection of ever saying no to his mother, not once in his entire lifetime. He couldn't recall ever disagreeing with her, or even saying, "You know, Mother, I see the situation differently." "Did you *ever* differ with her or voice a dissenting opinion?" I'd ask him. "No," he'd reply. "One didn't do that. There was no point."

Perhaps speaking in his own voice would have constituted an act of disloyalty and betrayal that carried with it an associated risk of abandonment. He watched his big sister defy their mother's authority and become irreparably and bitterly estranged from her. The relationships severed in the new country shared a history of profound grief and loss that was never processed or even mentioned. Perhaps sensing the emotional intensity that seethes when one family member refuses to acknowledge another, no one ever asked my grandmother about her family of origin or inquired about a past she never mentioned.

An Impossible Triangle

When my father married, the plot thickened. Rose, who had a pleasant disposition and got along well with everyone, quickly came to hate Archie's mother. She blamed her for ruining Archie's character, for spoiling him, and for being impossibly controlling and overbearing. As the tension mounted between the two most important women in my father's life, he had no idea how to navigate his part in the triangle. His mother said, "Send money!" Rose said, "We're not sending money. We don't have enough for our own family." His mother said, "You must move your family to California so you can help me out." Rose said, "We're not leaving Brooklyn." Archie, who couldn't speak up to even one woman in the simplest of circumstances, was caught in a loyalty struggle that paralyzed him. In the midst of this intense emotional field, he was unable to clarify what he believed and where he stood. The idea of alienating either his mother or his wife was intolerable, and he felt he had to choose between them. He "solved" the problem by saying yes to both (or at least by not saying no) and then behaving in sneaky and secretive ways in an attempt to appease each of them.

For example, when our family visited his mother in Los Angeles, she insisted that Archie look for a job there because she wanted him with her on the West Coast. Every morning he put on his good clothes and told her he was going out to look for work. Actually, he took a bus to the park and returned in the evening. My mother was so upset by this charade and so enraged at her mother-in-law, whom she blamed for Archie's behavior, that she packed her suitcase and flew to Seattle to stay with a friend. I was only four years old at the time, and I remember the terrible tension. Acting out rather than speaking out became a pattern for Archie. He defied both of the women in his life without standing up directly to either one. And he lost his self-respect along the way.

THE COST OF SELF-BETRAYAL: HIS AND HERS

Relationships require give-and-take, but a problem arises when one partner does more than his fair share of giving in and going along in marriage or family life. When it's the woman who assumes the accommodating position, she may suffer deeply and end up in a therapist's office, saying, "What's wrong with me?" (rather than "What's wrong with this relationship?"). But she does not bring into question what it means to be a woman. She's doing what the culture teaches, even if in modern times it may make her look like a caricature of femininity. For this reason, the price of silence and self-betrayal, of not speaking, acting, or thinking clearly, of not standing one's ground—these acts of self-sacrifice may not be keenly felt in the moment. As the peacemaker, the accommodator, the steadier of rocked boats, she is simply doing what women have always done.

Over time, however, the costs are dear. When a woman loses her resolve to speak up and stand firmly behind her position, she may be vulnerable to depression, anxiety, headaches, chronic anger, and bitterness. Sometimes these symptoms reflect an unconscious search for truth, forcing a more honest self-appraisal, including the degree to which she is voicing her authentic values and desires and living in accord with them. As writer Kat Duff puts it, "Sometimes I think we would lose ourselves altogether if it were not for our stubborn, irrepressible symptoms, calling us, requiring us, to recollect ourselves, to reorient ourselves to life." We need to listen carefully to the wisdom of our symptoms and to try to decode their meaning, because some of us have learned to settle, to fall silent, to deny that unfair circumstances exist or matter, and then to call our compromises "life." But our bodies, our deeper unconscious selves, remain harder to fool.

In contrast to the silent, accommodating woman, a man who feels powerless to use his voice violates our very definition of what it means to be a man. Consequently, he may then seek to prove his manhood in the most problematic ways: by being tough and aggressive, by acting up and acting out, or by removing himself emotionally from his relationships. He may be in a relationship where no one is going to tell him what to do, meaning he won't allow himself to be influenced or even moved by his partner. These are common male responses to feeling utterly helpless to right things through conversation or to speak with clarity, strength, and resolve.

My father's loss of voice was extreme, but from that vantage point we can see the ordinary in bold relief. We can surely identify with the dilemmas he faced. We may accommodate rather than negotiate with a partner or family member, telling ourselves, "It's not worth the fight." We may wall off parts of the self that we don't feel comfortable bringing into a relationship. We may protect ourselves by choosing distance over authentic connection. We may be caught in an intense family triangle. We may be reeling from hidden losses. We may be affected by unprocessed grief that no one talks about, including grief from our ancestors' immigrations or other traumatic dislocations. Finally, we may be hard-pressed to unlearn the lessons about silence and speaking out that we gleaned from our family of origin—the source of our first blueprint for navigating relationships.

CHAPTER 3

Our First Family: Where We Learned (Not) to Speak

Before all else, we are daughters or sons. Our relationships in our first family are the most influential in our lives, and they are never simple. Here, we first discover what thoughts and feelings we can say out loud, or even claim as our own.

My father's family obviously ranked rather high on dysfunction when it came to promoting the voice of individual family members. Even as adults, he and his siblings never gathered the courage to ask their mother a straightforward question—"Mom, do you have any sisters or brothers?" or "Can you tell me the names of your parents?" Reacting to the anxious climate of family life, they blunted their curiosity, narrowed their perceptions, and followed the "Don't ask, don't tell" policy that ruled the family. Children know at a deep, automatic level what they are not supposed to say or tell or even remember. Their utter emotional and economic dependency necessitates a fierce, unconscious loyalty to unspoken family rules.

As adults they may remain locked in silence, or attempt to belong by constructing a pseudo-self.

We all have a notion of the perfect family—a safe haven of unconditional love where we can speak our deepest truths and other family members will listen attentively with perfect empathy and attunement. Let's take a close look at what such a family looks like, at least in theory. Considering the ideal gives us the opportunity to feel even more terrible than we already do about the real-life family that fate dealt us.

A GLIMPSE INTO THE IDEAL FAMILY

The ideal family encourages the unfolding of each family member's true, authentic voice, promoting a sense of unity and belonging (the "we"), while respecting the separateness and differences of individual members (the "I"). Parents calmly enforce rules that guide a child's behavior, but they don't attempt to regulate the child's emotions or ideas. In this way, they create a safe space where kids can feel free to speak and be themselves.

Family members are comfortable sharing honest thoughts and feelings on even the most emotionally laden subjects without getting nervous about differences. Information flows freely, different points of view are respected, and difficult issues are discussed frankly. The emotional climate of family life is warm, spontaneous, and relaxed, so that children feel free to ask direct questions over time about whatever concerns them. Kids trust their parents to tell them the truth about important matters or, when appropriate, to say that some things are private and will not be shared. Children are seen objectively for who they really are, not through the distorted lens of what a parent wishes, fears, or needs them to be.

The parents are richly connected to each other and to their own families of origin, and together they model a vibrant, equal partnership in which conflict can be creatively addressed and resolved. Both parents can speak their minds and resolve their differences. Every now and then there's a good blowup (only dysfunctional parents never fight), but afterward the adults get their reactivity in check and offer a heartfelt apology when that's appropriate. No family member has to deny or silence an important aspect of the self in order to belong and be heard.

Since I'm describing one hypothetical "ideal" here, why not add that the universe will surely smile on such a family and bless it with a large measure of good luck? Nothing really bad ever happens to anyone. Or, if something bad does happen, family members pull together, calmly evaluate the facts, and then draw on their abundant resources from both within and outside the family to manage the crisis and turn it into a positive growth experience.

ENTER REALITY

You will feel relieved to know that this perfect family doesn't exist. In my many years of clinical practice, I haven't met the family that even begins to fit this description, at least not all the time. Of course, I haven't met every family. But I do know that the family is a sensitive system, reacting to the predictable stresses of the life cycle (such as the birth and launching of children) and to unanticipated stresses (such as chronic illness, untimely loss, and unemployment) that impinge on it. Also, many painful things have happened in the history of a family long before we enter the scene, and when issues are unresolved in one generation, they are often reenacted in the next. Finally, the powerful forces of racism, poverty, homophobia, and gender inequality deeply affect the interior of family life.

In real life, no parent can create the perfect emotional climate, like a garden greenhouse, for all conversation and for the cultivation of our authentic voice. Every family has its hot issues that can't be talked about productively—or even mentioned—because of current tensions or past history. Parents also project a great deal, confusing their children with themselves and with other family members. A projection becomes a story ("Kevin is always irresponsible, just like his dad"), then a prescription, and finally a self-fulfilling prophecy. Kevin may indeed have difficulty behaving responsibly, but labeling him ("the irresponsible one") and focusing intensely on his problem may lock Kevin into becoming the narrow story that is told about him, edging out other stories and possibilities.

One woman I saw in therapy was "the happy girl" in her family, a role that was rigidly enforced. When she showed any sign of sadness, her mother would say, "Who is this unhappy girl? That's not my Bea! My Bea has a pretty smile on her face! Let's make this sad little girl go away so that my *real* Bea can come back." Bea's mother was terrified of seeing any signs of depression in her daughter, because her own mother had been gripped by a consuming grief that she never got past. When Bea was visibly down, her mother made desperate efforts to cheer her up and to get her to "think happy thoughts." As an adult, Bea faced the challenge of sharing her feelings of sadness and depression with people who cared about her, rather than silencing and suppressing these important aspects of her experience.

Pushing the Polarities

Parents don't set out to constrict a child's voice or sense of possibilities. But when anxiety is high enough and lasts long enough, even the most resourceful adult can act badly. Parents overreact or underreact, distance or hover, say too much or not enough. When you observe any system under chronic stress, you'll see the extremes: the parents are rigid and authoritarian, or the family operates like a blob of protoplasm without clear leadership and boundaries. The lines of communication shut down, or everything is spilling over and kids aren't protected enough from adult anxieties. The parents operate as if they share a common brain and bloodline, or they're angrily polarized and can't reach a consensus about how to handle their difficult kid. The marriage gets stuck in too much distance or too much intensity. I could go on, but my point is that anxiety pushes the polarities and drives us to extremes, making it difficult for family members to talk calmly about any subject other than sports and the weather—and not always even sports.

Some folks are lucky to be born into a family where both parents have a high level of maturity, close ties to their own families and to each other, and lots of good fortune. This helps to create a calm emotional field in which family members can speak frankly. But any family can become dysfunctional (for want of a better word) if enough bad luck hits, and if the family lacks a circle of loving people to surround it, along with necessary economic and social supports. Many stressful things happen in family life that erode connection, block authentic engagement, and make it difficult for family members to communicate with spontaneity and vitality—or to communicate at all. When the emotional climate is sufficiently tense, even the dailiness of being oneself may feel impossible, let alone the challenge of togetherness.

THE COMPLEXITY OF REAL FAMILIES

Perhaps you're feeling a bit depressed now, because on the 1-to-10 rating scale of family health, the family you grew up in rates only about a 2 or 3 at best. Maybe you had little room to express yourself. Well, cheer up! It may help you to keep in mind that other families whom you are quite certain rank a 9 or 10 probably look a lot better from the outside. I tend to agree with author Mary Karr, who defines a dysfunctional family as "any family with more than one person in it." It's all a matter of more or less—although, admittedly, more or less can make a big difference.

Actually, I suggest that you avoid rating scales entirely. They can't do justice to our experience; real life is complicated, messy, unquantifiable, contextual, full of paradoxes and contradictions. My father, for example, was about as emotionally shut down as a person could be. But the wonderful sense of humor and totally unflappable manner (the upside of denial) that he managed to emerge with from his own intense family contributed to an often relaxed and irrepressibly silly emotional climate in our home. As my sister Susan said at his memorial service, "We all have so many happy memories of good times when the four of us sat around the dinner table telling jokes and stories, and laughing until we had tears running down our faces. A dinner was a real success if our mother could be reduced to total, speechless hysteria."

Conversation flowed freely around our table in Brooklyn. Okay, so we never mentioned my mother's cancer diagnosis or anything like that. But we weren't uptight. Susan (now a scientist) would regale the family with, say, little-known details about dissecting a frog, or we would listen spellbound as she told us the fascinating story about a particular whale whose six-foot-long penis had a bone that could be used as a cane. This was during the repressive silence of the 1950s, but we were a progressive family. I still vividly recall the shocked silence that occurred at a friend's house when, as a fifth-grader, I repeated my sister's interesting stories around their dinner table. I was not invited back.

Archie could also take a joke, even a bad one. One particular April Fool's Day, when Susan and I were ten and five years old, respectively, we awoke early and put salt in the sugar jar. Our father spit his first sip of morning coffee back into his cup with such vehemence that my mother thought he had gone mad. So she took a sip from his cup and reacted precisely the same, which left the four of us splitting our sides as our little trick was revealed. The ready accessibility of humor and bantering in our family made it a bit easier for me to open up some difficult conversations with Archie when, as an adult, I made it a project to challenge the long legacy of silence and distance in our relationship.

Every family member has strengths and vulnerabilities. So does every family form—single-parent families, stepfamilies, gay and lesbian families. Families, and the people who compose them, are always far more complex than the stories we can tell about them and the labels we use to describe them.

WHAT'S YOUR FAMILY'S HERITAGE?

Families from different cultural and ethnic backgrounds tend to have different conversational taboos. We all have a family legacy that has evolved over many generations, which dictates whether it's good or bad to boast, complain, ask for help, protest loudly, get emotional, take center stage, put ourselves first, or forgive somebody who's wronged us.

It's interesting to think about how our cultural heritage shapes our voices for both better and worse. Consider, for example, the Anglo-Saxon Protestant emphasis on optimism, cheerful stoicism, self-reliance, individual initiative, and problem-solving. As family therapist Monica McGoldrick notes, such family members may do fine as long as pushing forward and drawing on their individual competence and initiative gets them where they want to go. But these same strengths become a liability when a crisis arises that can't be "fixed," such as a chronic illness or an untimely loss. In such a situation, family members may have difficulty voicing their pain and vulnerability, accepting help, recognizing their interdependency, and supporting each other through the grief process. A no-muss-no-fuss attitude about emotionality may keep family members isolated, silent, and frozen in time.

In contrast, consider the high value placed on togetherness, emotional expressiveness, and "taking care of one's own" that characterizes many Italian-American families. Family loyalty and connected-ness are proud and admirable traditions. But let's see how these same forces can make it difficult for a family member to voice her limits and boundaries.

Family Loyalty: Too Much of a Good Thing

Maria, a homemaker with two children, had just celebrated her for-tieth birthday when her husband moved his elderly mother into their home. His work often kept him on the road, and his large Italian-American family had geographically dispersed, leaving Maria as the primary if not sole caretaker for his mother. Maria began to absorb more anxiety than she could manage, but since she thought she should be able to handle it, she didn't feel entitled to protest. Fur-thermore, she believed that her husband would not be able to tell his mother that the current arrangement was unworkable, and she could not even consider putting him in such an impossible position.

As Maria felt increasingly worn down, she began to complain to her husband. But she complained ineffectively, in a manner that invited him to disregard her concerns. As she found herself fulfilling the stereotype of the bitching, nagging, and complaining wife, she felt even more inadequate and incompetent. In actuality, she was protecting her husband by failing to take a firm, bottom-line posi-tion on her own behalf. She didn't say to him, "I love your mother, but I'm feeling so tired and depleted that I can't continue taking care of her in our home." Or, "I want to be helpful to your mother, but I can't be helpful all by myself. We need to call in the troops and consider other options."

Maria's mother-in-law eventually did move into an assisted-living situation, but only after Maria herself had a brief psychiatric hospitalization, resulting from the stress of being the primary care-taker. During our first therapy session, she said to me, "Thank God the psychiatrist who admitted me to the hospital explained to my husband that I can't take care of his mother anymore. I could never have said those words myself." Of course, she did find a way to let her husband and mother-in-law know that her caretaking limits had been exceeded. But this strong woman had been able to set her boundaries only by losing them—that is, by making a brief foray into mental illness herself. In a family where elderly parents had *always* been cared for at home by the woman, Maria was not able to say directly to her husband, "I can no longer do this."

Is It a Sin to Boast and Shine?

It's useful to think about our diverse family traditions, even on light matters that come up in daily conversation. Otherwise, it's easy to get angry at someone whose speech or silence is not to our liking. For example, I have two ambitious friends with remarkable achievements who deliberately downplay their accomplishments in social situations. In general, it irritates me when women hide their light under a bushel basket or act apologetic about their accomplishments ("Yes, it's an honor to be a Nobel Prize laureate, but I was just very lucky, and anyway, what I really love best is staying home and being a mom to Jake and Annie").

It's not that I want my friends to be big braggarts, but I do want them to claim achievement, admit to ambition, and share the facts. I think doing so is good for them and for the world overall. It exacts a toll, however imperceptible, on the human soul when we can't openly voice pride in our achievements, successes, and work well done. Or when we're complicitous with the disastrous feminine prescription to protect and bolster others by denying our ambition and hiding our ability. Plus, humility can become its own form of arrogance.

When I feel irritated at these friends, it helps me to remember that our different ethnic backgrounds are at play in this tension. One friend's Anglo-Saxon Protestant family thinks it's sinful to boast, even about distinguished ancestors. Standing out and showing off were heartily disapproved of by her parents, who encouraged her to become competent in a quiet way, never drawing attention to herself. My other friend tells me that "having a swelled head" was the biggest taboo in her Irish family. In contrast, my Jewish family considered it sinful for children *not* to give their parents something to boast about, and Susan and I were encouraged to shine. Hitting the winning home run was a far more important family value than being the good team player that my friends were supposed to be.

My father never hesitated to brag about his daughters' accomplishments to anyone who would listen, and he shamelessly embellished our achievements if the actual material wasn't up to standard. I'm reminded of the story of the Jewish mother Mrs. Kovarsky pushing a stroller with her two little boys in it. "Good morning," says a bystander. "Such darling little boys! So how old are they?"

"The doctor," said Mrs. Kovarsky, "is three, and the lawyer is two."

My father had only daughters, but the story has a familiar ring.

You might want to do your homework and find out where your ancestors emigrated from, how old they were, who they left behind, and how they managed. The more rooted you are in your history, the easier it is to operate from an authentic center and to feel more self-accepting. Playwright Marcia Cebulska tells me how relieved she felt when she visited her relatives in Poland and discovered a whole country of people who gestured as wildly as she did. Her relatives cried when they said good-night and greeted her in the morning with smiles and songs. "My own emotionality received validation," she explains. "After that I tried less hard to be WASP-like, and I stopped sitting on my hands to keep from gesturing. It still helps to mentally revisit these people in the old country."

Generalizations obviously run the risk of simplifying and stereotyping people, and keeping them down. There's great diversity in any group and countless exceptions to every rule. Also, many of us don't know what box to check when we're asked to classify the origins of our ancestors, because we're the products of complex multiethnic or multiracial traditions. But if we're not knowledgeable about the cultures and traditions of different people, we're bound to ascribe too much pathology to our own family—or to that other person or group we think is saying too much, too little, or the wrong thing.

The Power of the Dominant Culture

Another meaning of *culture* is "the dominant culture" which shifts with the economic, social, and political climate of the day. The dominant group puts some subjects, and some families' experience, in the realm of the unspeakable, or simply the not spoken, or the not heard, or the not counted.

Many subjects that are common conversation today were shrouded in silence when I was growing up in the 1950s. The unspeakable back then included a diverse range of facts, such as a parent's cancer diagnosis, the adoption of a child by an infertile couple, the love between two women, a baby born to a single mother, or the violence happening in somebody's home. The unspeakable also included such feelings as a mother's rage at her child or her desire for an ambitious, adventuresome life outside the confines of her prescribed role.

These particular examples of secrecy and silence would be unusual today because there is more information and less stigma surrounding these particular matters. Thank goodness for social advance, as the civil rights, feminist, gay and lesbian rights, and adoption reform movements. The history of such movements shows us how people can alter the narrow and stigmatized meanings that the dominant culture assigns to certain subjects or groups.

As new meanings develop and become established, more individuals come forth to tell their truth, to offer more accurate accounts of their lives. Previously dishonored individuals, families, and communities begin to develop a proud identity based on an authentic voice and the shedding of silence, secrecy, stigma, and shame. When we collectively challenge the shaming and stigmatizing myths of the dominant culture, we make room for more honest conversation in the deepest interior of family life.

THE FAMILY FATE DEALT US

I won't even try to tie things together here with pat conclusions such as, "If you grew up in *this* sort of family, the development of your voice will be affected in *these* ways, and here are six steps to fix the problem." Such formulas don't begin to do justice to the complexity of human experience and to the unique ways children respond to their families.

Our family is the deck of cards fate handed us, giving us virtually no say whatsoever about where we landed. If there's any justice, it's that our parents didn't get to handpick us either. They didn't have a clue what they were signing up for when they brought us into the world or adopted us, and they, too, are stuck with what they got. And of course our parents also had to contend with a history of their own that was not of their choosing.

But here's the most reassuring fact of all. There's something to be said for being part of a real, human, flawed family. Painful family relationships are part of the stuff of growing up, and even bad experiences foster learning. If I had to account for my choice of profession as a psychologist and writer, and the clarity of my voice (when it's clear), I'd have to give as much credit to the more wrenching details of my childhood as I would to my family's enormous resilience and strength. It's not just that we look around at the secrets and silences of our first family and learn what *not* to do. It's also that the human spirit is remarkable, unpredictable, and lawless. Many children respond to the painful silences or intensities of their past by developing great strength of character or a special gift to speak courageously and well.

Working to restore our voice with members of our first family can be a terrific learning experience. We didn't choose these difficult folks, but, as adults, how we talk to them is up to us. Observing and changing our part in family conversations is one royal road to change. In other words, if you can learn to speak clearly and to respond in a new way with your difficult mother or sister, then other relationships will be a piece of cake.

Finally, we don't have one true voice that might have bloomed unfettered and free had our family not screwed us up. It's important to remember that the self is continually reinvented through our interactions with others. Every relationship is a laboratory in which we can practice using our voice in new ways and observe the results of our experiments. Some of us need to practice voicing our strength. Some of us need to practice voicing vulnerability. It doesn't matter where we begin.

CHAPTER 4

Should You Share Vulnerability?

I'm a fairly open person, and I've become more so with age. Getting older brings the comforting knowledge that the things we consider most shameful and weird about ourselves are actually pretty universal—or if not, that other folks have their own shameful and weird stuff. This growing realization that we're not so unique makes it easier to share who we really are and how we got there.

Recently, I found my lock-and-key diary from sixth grade and read from it to my trusted friends Emily and Julie over coffee. The previous week, we had agreed that we'd each bring a sample of our writing or something else revealing from the past to share with each other—a sort of adult show-and-tell. In the particular diary I brought to the coffee shop, the sentence "I love Michael Howard Sacher" covers the pages from July through August and is repeated (I noted in my diary) 961 times. In the front pages are pasted a shoelace from Michael's sneaker, a bobby pin he had bent and used to snap me with, a spelling paper and a composition I must have taken from his desk, a clothes tag from his underpants (the story about how I got this is too long to tell), and more. Then there are stellar literary passages, such as the one I read aloud to my friends:

> After school me and Marla got into some snowball fight with Eddie B. and Jonny L. Jonny washed my face four times with snow and threw snow in my mouth, ears, up my skirt and got my hair so full of snow it wasn't funny. He made me get down on my hands and put my head in the snow and he put his feet in my hair. He made me beg like a dog. Oh, why couldn't it have been you Michael. I love you forever.

So I was a teensy bit embarrassed to be sharing such prose with my friends, especially since I followed Emily, who read journal passages that were enviably deep, lyrical, and indicative of an early feminist consciousness. But then again, when you reach my age, your life is no longer too embarrassing because you realize that *everyone's* life is embarrassing. This revelation allows a certain freedom to be oneself.

I kept a lock-and-key diary for almost all my growing-up years in Brooklyn, and not one entry shows any indication of literary talent, insight, imagination, or courage. Today, when I'm invited to talk to students in the public schools about writing, I bring a diary along and pass it around. If, say, I'm going into a ninth-grade class, I bring my ninth-grade diary. The students are enthralled. They thumb through the pages and say to one another, "Wow! Can you believe that *she* writes books?!" My diaries—which give new meaning to the word *shallow*—bolster their confidence, teaching them that writers don't have fairy dust sprinkled on them but rather are just plain folks. We help others as much by sharing our only-too-human side as we do by sharing our skills and competence.

★   ★   ★

Of course, there are different kinds of revelations and self-disclosures. In certain ways I've become more private as I mature and move through the life cycle. When I first met my husband, Steve, we were graduate students, and I kept him up into the wee hours of the morning regaling him with tales of my previous erotic adventures. Telling all was my style. It suited us both fine and drew us closer together. I wanted him to know everything about me, just as I wanted to know everything about him.

But were I to embark on a new love relationship today, in my fifties, I can't imagine sharing such details, or even remembering them. I've become more private, my history too long to run through, and I consider some information sacred. In midlife, I make more careful decisions about whom to tell, what to tell, when to tell, how much to tell, and how to tell. Yet I am more "myself" than ever before. With age comes more discernment, more knowledge of the self, and less willingness to compromise or betray the self to keep relationships calm—or to keep them at all. The recognition at midlife that the future doesn't stretch out forever challenges us to figure out what really matters and to speak wisely.

How Much Should You Reveal?

My friend Jennifer Berman once drew a cartoon depicting a woman approaching a man at a party. She's saying, "Hi! My name's Gloria. Allow me to tell you all the heart-wrenching details about my tragic childhood." In our desperate rush to become intimate, we may tell too much too soon. Sharing vulnerability is one way we feel close to each other, but sharing indiscriminately or prematurely has the opposite effect. If you meet someone at a cocktail party who discloses her most searing pain standing over the chopped liver, you'll probably question her judgment and maturity rather than admiring her openness.

We create a healthy boundary around the self by exercising some control over what we conceal from—or reveal to—others. When it comes to sharing vulnerability, it's wise to take time to test whether the other person is worthy of hearing our stories and to assess our own level of safety and comfort in sharing sensitive material. We want to trust that the other person isn't going to deny and minimize our pain, or alternatively, overfocus on our problem in an unhelpful way. We don't want to be put down, pitied, or gossiped about, nor do we want to have sensitive information used against us.

Not everyone finds solace in revealing the personal. People have different coping styles, and some individuals are more private than others. While self-disclosing is one way to be intimate, social psychologist Carol Tavris reminds us it's not the only way. She writes:

> Years ago, my husband had to have some worrisome medical tests, and the night before he was to go to the hospital we went to dinner with one of his best friends who was visiting from England. I watched, fascinated, as male stoicism combined with English reserve produced a decidedly unfemale-like encounter. They laughed, they told stories, they argued about movies, they reminisced. Neither mentioned the hospital, their worries or their affection for each other. They didn't need to.

Tavris reminds us that love is communicated in different ways and that connections take many different forms. Maintaining privacy isn't just a way to hide out; it can simply be our preferred way of being.

Whether we're open or private, we all seek to control the flow of personal information about ourselves. My friend Alice had chronic back pain for over a year until her problem was corrected by surgery. She told me that people were constantly asking how she felt—whether or not she welcomed their inquiries. Alice recognized their good intentions, but on most days she preferred not to talk or even think about her problem. A part of her life that she wanted to keep in the background became foreground, with each inquiry reminding her that her pain was always visible to others. Of course, she was in control of her response to unwanted questions and how much she chose to reveal. But she experienced a loss of privacy that many of us take for granted—the ability to protect ourselves from unwanted attention and encroachment, buffered by a certain amount of emotional space that we take to be "ours." As Alice put it, "I lost control over the conversation."

Thank goodness for that remarkable human capacity to show the false and hide the real. We say, "Fine," in response to the question, "How are you?" if we don't feel like mentioning that our house just burned down and our son just dropped out of his drug treatment program. Even with an intimate partner, sometimes we want to keep our feelings to ourselves or to pretend during dinner that we're in a good mood when we're not.

Whenever we feel acutely vulnerable, the first line of business is to do whatever it takes to feel a little bit better. This may mean baring our soul to a trustworthy party who can give us the gift of empathy and attention. It may mean talking to someone who will either help us make a plan or even take over for us for a while. Or it may mean going to a movie, reading, gardening, or walking alone in the woods.

I have a colleague who forces herself to smile as part of her spiritual practice, even when she's feeling depressed. She quotes Thich Nhat Hanh, a world-renowned spiritual leader and peace activist who encourages us to smile often. He notes that the act of smiling relaxes the muscles of the face and has many benefits.

Putting on a happy face is hardly useful, however, if we habitually conceal real feelings that need to be acknowledged, shared, validated, and understood. And no amount of acting as if things are okay can substitute for connections to caring people with whom we can really be ourselves. But if sharing our pain digs us deeper into it, and pretending to be happy in a particular situation empowers us to feel better and act on our own behalf, then pretending may be the first order of business, or at least a reasonable first step.

When it comes to voicing vulnerability, we may be full of contradictions, open in some ways yet closed in others. For example, one friend of mine is the most "out there" person I know, in terms of being lesbian. She has never once, no matter what the circumstances, tried to pass for straight. If a woman next to her on a bus asked, "Are you married?" she'd say, "Well, my partner, Cynthia, is a woman, and we think of ourselves as married." She has no tolerance for hiding, even through silence about who she is and whom she loves. She confronts injustice forcefully, and she is boldly and courageously herself in refusing to hide the affections of her heart. Her voice, in this important regard—and on all matters political—is unflinchingly honest and brave.

But this very same woman has enormous difficulty sharing her feelings of vulnerability with anyone close to her. A real do-it-yourselfer, she rarely acknowledges her own need for help and support. While she intellectually believes in the healing power of confiding in others, she herself is no good at it. As the eldest child of alcoholic parents, she had no experience of voicing her emotional needs and having them met. As an adult, she gains deep satisfaction from her capacity to give generously and to take care of others, but she is profoundly guarded against letting anyone return the favor. When she does share a serious problem, it's as if she's fiercely sweeping the ground in front of her to keep the other person from getting near her or emotionally connecting with her pain.

Over time, our self-regard and our ability to be intimate suffer when we are unable to put forth both competence and vulnerability in a balanced way. Our silence may protect us from fear in the short run but leave us more shamed, isolated, and alone in the long run, without the necessary emotional and practical support we need and deserve. The more intimate and enduring the relationship, the greater the longing to find some way to share our full selves, and the greater the consequences of not telling, of not being real. When a crisis strikes, it will hit us especially hard if we haven't had some practice in letting people know how much we need them, and in accepting the help and comfort they are able to offer.

IS IT A SHAME TO BE SICK?

Consider Pamela, who came to therapy two years after she was diagnosed with a chronic progressive illness. When I first saw her, she felt compelled to appear plucky and upbeat, not only because she was convinced that "negative emotions" would threaten her already compromised immune system, but also because she was terrified of alienating people, especially her partner, with her grief and pain. She told me that her worst fears were of losing control and being dependent, which was the direction her illness would likely take her.

Obviously, it's not useful to drown in despair. Any of us can get so focused on the negative that we can lose touch with our strengths, including our ability to manage hardship and our capacity for optimism, hope, and joy. But Pamela's belief that she had to maintain a mood of cheerfulness was wearing on her. It took energy to conceal and deny her real emotions and to pretend to her friends and family that she could always put aside her fear. It shut down the lines of communication and the possibility of a deeper intimacy with her partner.

Pamela was locked into a position of unutterable loneliness until her partner, Sam, joined us in the therapy process. In this safe space, they learned to talk openly together rather than to continue "protecting" each other from painful facts and feelings. For Pamela, the hardest emotional piece was the shame she felt at being ill and dependent, and even for simply having less energy and ability than she did before. Over time, she was able to give voice to her shame, expose it to the light of day, confront it, and cut it down to size. She could not exorcise it entirely from her day-to-day experience, but she learned to greet shame—like fear—as an uninvited visitor who could be counted on to come and go. And she didn't let shame stop her from giving voice to her needs and accepting comfort and help.

Sam also learned to give voice to his vulnerability. He learned to say, "There are limits to what I can do." He needed to pay attention to when he was absorbing more stress than he could handle, so that he could take better care of himself. This was a huge challenge for Sam, who tended to keep functioning until he was physically exhausted and totally depleted. So Sam, too, became more expert at sharing vulnerability, at letting Pamela as well as their friends and family know when he needed help and support for himself. Sometimes he blamed Pamela for being too demanding, because she wanted *only* him to do certain things for her. But it was ultimately his responsibility to clarify what he could and couldn't do.

When one person in a relationship has special needs, learning to give voice to our vulnerability is a challenge for both parties. Whether we're dealing with a parent, a sibling, or a dear friend, it's never easy to sort out the limits of what we can give or do—especially if the other person wants us (not someone else) to be there. It can be hard to say, "I love you, but we are in a place where we both need more support. Acknowledging this is painful for both of us. I know how painful it is for you to go through this. And it's painful watching you go through this because I love you so much. But we need to get other people involved, because I can't do it all by myself."

Learning How Not to Be Helpful

Sam also learned to sit with Pamela's expression of grief, to give her the gift of his full attention without trying to fix her feelings or shower her with advice. He was scared that if he really made room for Pamela's grief, and for his own, they would lose their capacity for joy and hope. Of course, the opposite was true. Feelings come as a package deal. We can't deny our rage, pain, and vulnerability without also denying our capacity for joy, love, and intimacy. Nor does denying grief work well for most people in the long run.

Learning to *not* give advice, and to ask questions instead ("What's your greatest fear about how your illness will affect us?") was no small achievement for Sam, because it went against his automatic way of operating. When I first met Sam, he thought it was his sacred calling to fix others, and he didn't pay attention to whether his advice was actually appreciated at a particular moment. When Pamela was feeling emotionally flattened, he was quick to editorialize about her situation and to tell her what she should do to improve her attitude and health. Or he might give her a pep talk or a spiritual boost. He felt that he had to have some kind of answer, even when there was no answer to be had.

Sam made an important step forward when he learned how *not* to be helpful, except when Pamela was asking for help. More accurately, he learned that when Pamela was grieving, the best way to be helpful was to ask questions and offer her his pure attention and love. When he was uncertain about what Pamela would find most supportive, he learned to ask her. Did she want him to simply listen and ask questions? Did she want to hear more about his own feelings about her situation? Did she want advice or help making a plan? Would she feel reassured if Sam expressed confidence in her ability to manage her illness over time—or would she feel he was offering false reassurance and minimizing her struggle? Would she like him to go rent a video to distract her from feeling down?

Sam also learned to cry with her, a remarkable achievement, especially since he had the false notion that he was "protecting" her by hiding his own grief and "being strong" for her when she was feeling weak. They learned that grieving together did not cause them to lose their capacity for strength. Instead, it brought them closer together.

There is a "voice lesson" here for all of us. In many circumstances, the most helpful step we can take is to *not* be helpful. Instead, we can usually be more supportive to distressed persons we love simply by caring about them—by being emotionally present without pulling back from their pain and without trying to take it away.

Rushing in to offer advice—or to cheer someone up—may reflect our own inability to remain emotionally present in the face of another person's problems and pain, or to experience our own. If we move in too quickly with solutions, we can make it harder for others to be in touch with their own competence and inner resources, and we unwittingly rob those we love of the opportunity to feel what they are feeling and express it to us. Learning to be an attentive, caring listener and a skilled questioner can empower others to search for their own solutions. It also helps us to get in touch with our own vulnerability, which paves the way for a richer intimacy between two people.

THE MYTH OF DEPENDENCY

Dependency has gotten a bad rap. (Can't you say "*inter*dependency"? a colleague wonders.) The fact is, we all depend on others. It's part of the human condition. We can pretend this isn't so when we're young and healthy and our work is going well. But even when things are moving along swimmingly, we all rely on multiple services, systems, and structures just to get through the day. We don't notice this fact when all goes well, as it typically does for middle-class folks. But let's say your computer crashes, your health insurance gets canceled, your best friend moves away, your car gets stolen, and you can't find a dentist in an emergency. When the systems that support you break down, you learn just how dependent you really are.

The cultural emphasis on independence is so strong that we may actually feel ashamed of our dependence. The shaming of people who have special emotional or physical needs runs so deep that we fail to question it, yet we need to. Writer Anne Finger puts it so well. "We have this notion that some dependencies are OK and others are not," she writes. "It's OK to need a car; it's not OK to need a wheelchair. It's OK to go to a hairstylist to get your hair done; it's not OK to need an attendant to wash your face and hands."

People may also feel shame for their honest suffering. Everyone suffers sometime, yet we're taught to tuck it away, to deny grief rather than welcome its expression. As bell hooks notes, we may feel shame especially about grief that lingers: "Like a stain on our clothes, it marks us as flawed, imperfect. To cling to grief, to desire its expression, is to be out of sync with modern life, where the hip do not get bogged down in mourning."

Surrendering to forces or events larger than we are isn't our cultural orientation. We have a constitutional right to pursue happiness (not truth or wisdom, mind you), and it's our job to "take control of our lives" and run the show. We're expected to turn even the most unfathomable losses into an opportunity for personal growth. Writer Michael Ventura calls this cultural expectation our "consumer attitude toward experience," and notes that other cultures find this attitude unnatural.

Dependency and suffering are essential components of the human condition. Sooner or later harsh experience teaches us how much we need each other. The only aspect of either that's really shameful is the persistent and false societal belief that people can bootstrap their way to health, wealth, and happiness. In fact, it's wise to get some experience in voicing your vulnerability, needs, and limits during calmer times, before the universe gives you a crash course on how much you really need people.

Is Dreading Dependency a "Guy Thing"?

It's a cultural cliché that men can't ask for directions, and certainly not for emotional help. Fortunately, that's changing. In the past, a man rarely came to therapy for a relationship problem unless his wife insisted or he was in the crisis of relationship loss. Today, some bold men are pioneering a new way.

Dan, a young man I saw for a brief consultation, accepted my suggestion that he embark on the adventure of therapy with a therapist in his hometown to work on his relationship with his father. In a note of thanks to me, Dan commented on his initial resistance to the idea:

> Men often feel life is some sort of sporting contest and that to receive "help" is to have your record rendered a bit less valid, like a "wind aided" long jump record. I've seen myself as an 80-year-old man sitting contentedly at a barber shop saying, "Never broken a bone and never been to a therapist." As if that meant climbing Everest without the oxygen tanks. Hopefully, I'm being permanently disabused of such notions.

It takes courage to ask for help. But it's not just men who are prone to shame attacks for voicing vulnerability and asking for assistance. Many women have this problem too, especially eldest daughters who learned growing up that they could not rely on a parent to be competent and nurturing. In this common circumstance, a woman may possess a towering competence but have enormous difficulty in putting forth the more sensitive, vulnerable parts of herself. It can be quite an enormous achievement for her to say to a partner, friend, or family member, "I'm having a terrible day. May I come over and talk?" Or, "I'm exhausted. I can't keep doing so much." Or even, "My house is out of control. I'd be so grateful if you had time over the weekend to help me get organized."

WHAT'S YOUR PERSONAL STYLE?

What are your core beliefs about sharing vulnerability? Do you believe that you honor the people in your inner circle by sharing your pain and allowing them into your experience? Or, alternatively, do you feel like you're dumping messy emotions on family or friends, making them uncomfortable, putting them on the spot, or spoiling their day? Do you think there is something superior and noble about people who don't complain, who hide their pain and quietly go about solving their own problems? Or, alternatively, do you believe that the strongest, most centered people have the capacity to be open and "out there" with the facts, and with the full range of their emotional experience?

In addition to our beliefs about self-disclosure, we all have patterned ways of managing anxiety and getting comfortable in our key relationships. Both Pamela and Sam, for example, were staunch *over*functioners before they had to accept chronic illness as their constant companion. This meant that they would both move in quickly to advise, rescue, and take over when stress hit. They had difficulty sharing their vulnerable side—especially with people they viewed as having problems of their own. Pamela and Sam were seen as "always reliable" and "always having it together," and they were both invested in maintaining this image without being aware of the cost. Before Pamela's illness, neither could give voice to both their competence and their vulnerability in a balanced way.

Other individuals are staunch *under*functioners. They become less competent under stress, inviting others to take over. They tend to become the focus of family gossip, worry, or concern, and they frequently earn labels such as "the fragile one," "the selfish one," "the problem child," "the irresponsible one." They may have difficulty showing their strong, competent side to intimate others, and they just can't seem to get organized.

Every person has both strengths and vulnerabilities; but when we're out of balance, it won't help to just "be ourselves" and "say what we feel," which boils down to doing more of the same. Instead, we may need to shift how we *present* ourselves, so that we can expand how we *experience* ourselves with others. We need to find ways to remind ourselves—or even trick ourselves into remembering—that we are each capable of enacting many different versions of ourselves.

Most of us find it hard to switch off automatic pilot, to alter our unproductive, habitual ways of talking to others. An old Spanish proverb reminds us, "Habits at first are silken threads, then they become cables." If you're an *over*functioner, you may not really believe that others have very much of value to offer you. You may need to practice sharing the more tender, vulnerable parts of yourself, and to stop having the answers for anyone else. If you're an *under*functioner, you may need to do just the opposite; that is, to tone down your expressions of vulnerability and to amplify your strength and competence, even when you don't think it's there. Whatever your patterned behaviors may be, you can cultivate a more authentic voice—if you're willing to practice doing what does not come naturally.

CHAPTER 5

In Praise of Pretending

Having an authentic voice doesn't mean that we say everything we think or express whatever we feel. Rather, it means that we can think about what we wish to express and what we hope to accomplish. Being our best selves may require us to exercise restraint, even to *pretend,* if necessary.

Pretending can be a creative venture. Pretending joy and courage, for example, can have the effect of a self-fulfilling prophecy, helping us discover or enhance our capacity for these positive feelings. Paradoxically, we can learn what is true, possible, or "still there" in ourselves and others by experimenting with new behaviors that may initially make us feel like we're not being our real selves.

A Whitewater Adventure

I sometimes voice vulnerability to a fault. An example that comes to mind is my first whitewater rafting trip.

In a moment of great courage or ignorance, I joined a group of colleagues conducting a seminar for business executives on the subject of human behavior. This particular week-long seminar was not being held on the familiar—and dry—campus of the Menninger Clinic, where I work, but instead would take the participants down the Yampa and Green Rivers in Utah and Colorado. A core Menninger staff group had been conducting river seminars for some time, with the cooperation of Colorado's Outward Bound program. Several of my colleagues had become veteran whitewater rafters, but this was a first for me.

I enlisted for the rafting trip without any prior camping or whitewater experience. My only experience on water was floating on placid lakes in rowboats and canoes, usually with someone else paddling. As luck would have it, on this trip drenching rains made it nearly impossible to stay warm and dry, with the river surging higher and faster than it had in years. I struggled to master everything, from tying gear securely into the raft to understanding what commands I should shout to guide the raft down the rapids when it was my turn to captain. I was the most frightened and least competent person on the river, a status magnified further by the fact that women were rare on this trip.

The more anxious I became, the more others focused on me with a critical or worried eye. I was so open about feeling frightened and incompetent that it became harder for the participants to see my areas of strength, even though I continued to be a skilled lecturer, consultant, and small-group leader, which was why I was there in the first place. Only midway through the seminar did I realize that I needed to act brave and use discretion in choosing confidants.

I turned this corner at a low point, before we ran the most difficult and unforgiving rapid. I had stayed up much of the night imagining my impending death, and when morning broke, I was a wreck. But I had resolved to try to act confident, and so I pulled aside just one colleague, Nolan Brohaugh, whose maturity and kindness I trusted, to confide in. I told him how completely terrified I was and instructed him on what he should tell my husband, Steve, and my boys in the event that the river took me. Nolan listened attentively and calmly shared his perspective and concerns. He didn't overreact or underreact. He didn't minimize my fears or offer reassurances. As we headed toward the dreaded rapid, he didn't patronize me, distance, or hover. He was warm and available, but he never treated me like "a problem," nor did he lose sight of my overall competence.

With everyone else, I forced myself to act brave. I had determined that it wasn't useful to let my fear spill over into the group. Indeed, when I began to feign courage and calm, it helped me to muster those very qualities in myself, and the people I was with began to treat me as a real member of the team rather than the weak link in a chain. I actually ended up having some of the best times of my life on the river—that is, when I wasn't having some of the worst.

Of course, work relationships are not the same as intimate relationships. The primary task in a workplace is to meet the requirements of the job, while the primary task of an intimate relationship is to deepen the experience of knowing the other person and being known. Still, an anxious work system isn't all that different from an anxious family system. In each setting, folks are looking for ways to soothe their own anxiety and focusing on a vulnerable member, as if she or he were the source of the problem, is one way to do that. As a friend of mine puts it, why raise your hand and volunteer for the job of scapegoat? If you tend to become the focus of family worry, gossip, or concern, you might want to experiment with sharing less of your vulnerability and instead display more of your strength and competence. (I promise you that it's really there, even if you're convinced it's not.)

There may be good reason for you to modify expressions of vulnerability in a couple relationship as well. You may be wearing the other person down, or deepening the negativity in your own brain. You may be contributing to an imbalance in the relationship, where your partner can disown his own vulnerability because you are doing the "feeling work" for two. You may unwittingly be moving the relationship in the very direction you don't want to see it go, and eroding your own self-esteem along the way.

CAN YOU STOP PURSUING?

Consider the familiar pattern of the pursuer and the distancer. She keeps pressuring him to commit, while he keeps hedging. Soon the pattern takes on a life of its own. All she can do is express dependency and neediness. All he can do is maintain a cool distance and assert his need for space. She's anxiously pursuing, and he's putting on his track shoes. The behavior of each provokes and maintains the behavior of the other.

Once this pattern is set in motion, I may suggest that the pursuer consider a bold act of pretending. Can she put aside a period—say two weeks—to take her focus off her partner and put her energy into her own life? Can she be the one to seek a bit of separateness, for example, by being less available and by going out more with her friends? Might she initiate a talk in which she shares her own doubts about commitment? (We all have them.) Can she back up—but still stay warmly connected to her boyfriend, since coldly retreating into the distance herself won't shift the pattern in a productive way? Can she begin to share her anxiety and vulnerability with her best friends rather than her partner?

Such advice may sound like the old play-hard-to-get tactic that women once were taught to use. If we want to develop an authentic voice, why be strategic in a way that smacks of those old phony and manipulative dating games? Because there's nothing authentic or real in compulsively continuing a pattern where one person only pursues and the other only distances. Interrupting this cycle enables each partner to begin sorting out the complex internal experience of both wanting and fearing intimacy.

Interrupting the pursuit cycle—or any habitual pattern—won't solve the relationship problems. Two people still face the challenge of figuring out whether they can find a mutually satisfying balance of separateness and togetherness. But when we can only voice more of the same (she wants to get hitched, he wants to stay free), we may stay stuck in a narrow view of what is true and what is possible. Experimenting with a different voice can teach us that what we think or feel isn't set in stone.

"SHE'S LEAVING ME!"

If our automatic tendency is to anxiously pursue someone who is wanting space, it's incredibly difficult to move away from that pattern of behavior. Nor is the role of pursuer the exclusive domain of women.

Tim came to see me after his wife of four years said she wanted "a temporary separation to think things through" and moved into her girlfriend's apartment. He suspected Jill of having an affair. When I first saw Tim, he was filled with self-recrimination. He described himself as "a total jackass" who had put all his energy and attention into his work and had ignored Jill's loneliness and frustration. A year earlier, he had insisted that they move into a "dream house," which he now realized wasn't *her* dream house, and their relationship deteriorated even more after that. Jill often fought with Tim, but she would ultimately accommodate rather than hold her ground, so it came as a shock to him when she questioned their marriage and moved out.

Tim was devastated. He called Jill frequently, crying on the phone, begging her to come home, and promising to make any changes she wanted. Jill responded by acting colder and distancing further. Tim then pursued her even more desperately. Jill, in turn, told him that she didn't want any more contact or conversation until she felt ready.

When Tim came to see me, he described an urgent need to "fight like hell to win Jill back." He was leaving several messages a day on her phone machine, sending her long e-mails, and asking her closest friends to intervene on his behalf. In our early sessions, Tim sounded like a broken record: "I have to see her. She can't shut me out like this. If I could just get her to come home for one day, I know I could convince her to stay."

When we feel acutely vulnerable, we need to do whatever it takes to calm our emotions. But here's the caveat. Like an addict, Tim felt driven to actions that soothed him only temporarily and drove Jill further away. It was clear that Jill had a visceral negative reaction to his frantic moves toward her and to his expressions of neediness and vulnerability. The more he pursued, the more she distanced. The more Tim conveyed, "Take care of me, I'm falling apart," the more Jill wanted out.

Yes, Tim needed to give voice to his pain, and he deserved to get all the help, love, and support he could garner during this crisis. But if his goal was to give his marriage the best chance of succeeding, he needed to move toward *other* people and start to take care of himself. To this end, I suggested that he talk with a man he liked at work who had recently gone through a painful divorce. I also encouraged him to reach out to a few family members who could support him. These were new ideas to Tim. Like many men, he had no experience confiding in friends, nor had he ever shared his personal problems with his parents or brothers. Tim's willingness to begin talking openly with people who could provide him with support, was a courageous act of change on his part.

After Tim calmed down, he was able to think strategically about how best to approach Jill. He composed a brief note that was a major departure in content and tone from his previous lengthy tomes. His words were thoughtful and self-focused.

> Dear Jill,
>
> I apologize for being on your back and not respecting your need for space. I feel like I've turned a corner. I've started therapy and I'm getting the help I need. This crisis has forced me to take a good, hard look at myself and my contribution to the problems in our marriage. I realize that I need to focus on my own issues at this time. I know we both have a lot to think about, and I support whatever you need to do for yourself. I want you to know that I'm going to be okay and that I'm taking good care of myself. I love you and I hope we can make the marriage work. Whenever you're ready to talk, let me know.
> Love,
> Tim

Tim knew that these words outlined what he should do, but he had no confidence he could stick with his plan. He still felt compelled to call Jill again and again, so his stated intention to do otherwise seemed false to him. But what at first felt like pretending led him to get more in touch with what he later called "the higher truth." He *did* need to work on himself. Jill *did* need space. He *was* going to be okay—even though it didn't feel that way to him at first. Tim's accomplishment was not simply that he penned the strategically correct words to Jill. More important, his words reflected a mature and loving attitude that he thought was out of his reach, and they summarized a plan that could move him toward becoming the person he wanted to be.

Tim's simple note to Jill was a remarkable achievement. You need to muster a huge amount of maturity and discipline to say less when you feel compelled to say more, to not pursue when you feel desperate to do so, and to keep communication focused entirely on yourself. This is especially true when what you really want to do is convince the other person to think, feel, or behave differently.

Swimming against the Emotional Tide

Perhaps nothing leaves us more vulnerable than the threat of relationship loss. When we're drowning in emotions, it's impossible to think creatively or clearly. We may *think* we're thinking, but in reality we're just reacting.

Thinking logically sometimes gets a bad rap because we confuse it with intellectualizing, which is a way of avoiding feelings. Men are often criticized for being logical instead of emotional, for problem-solving rather than focusing on feelings. A defensive overstriving for objectivity can lead us to close our hearts to both the suffering and the joy within and around us, and so it separates us from our full humanity.

But it's never useful to drown in emotions or to lose the capacity to think about how we want to express them. In Tim's situation, staying with his feelings—and voicing them to Jill—wasn't going to help him for more than a few minutes during his time of crisis. It was Tim's capacity to *think*—to consider how his words were influencing Jill and affecting his own well-being—that ultimately saw him through the crisis. His willingness to seek help and line up support for himself was a critical first step.

Tim and Jill did eventually get back together. His ability to change caught Jill's attention. His new behavior invited Jill to slow down, to consider her part in their problems, and to remember all that was good in their marriage. But it might not have happened that way. Try as we may, we can't control the outcome of a situation when another person is involved.

Most important, Tim was able to control his own reactivity, and he worked to resist the strong emotional currents that initially drove him. He continued to navigate his part of the process in a thoughtful way. His capacity to focus on changing himself put him on more solid ground—whether Jill returned or not.

GOOD PRETENDING/BAD PRETENDING

Obviously, not all pretending is bold and enhancing. We can pretend out of fear—out of a wish to please and be loved, or to cling to what we have at any cost. We can pretend in order to avoid a leap into the unknown. Even when we're miserable in a relationship, we may feel that the devil we know is better than the devil we don't know. So we may pretend—even to ourselves—because we don't want to introduce greater clarity into our relationship, and we're scared to use our voice to bring our knowledge of ourselves and our partner into sharper focus.

This kind of pretending ultimately makes our voice small. It limits our possibilities and potentialities rather than expanding them. Pretending can involve misguided acts of self-sacrifice, and grave, ongoing deceptions, shored up by lying and self-betrayal. Too much of the self (our wants, beliefs, priorities, values) disappears or becomes negotiable under relationship pressure. This is obviously not the kind of pretending I recommend.

In contrast, bold acts of pretending take us out of repetitive and narrow conversations into the territory of the unknown. In intimate relationships, pretending can facilitate an expanded self-awareness by making a dent in our habitual ways of responding to others. We can't always see what's true or possible in a relationship or in ourselves until *after* we change our behavior. And what begins as pretense, as experimenting, can lead in time to a more richly textured experience of our self and our partner. Pretending can enlighten us, lead us to invent and discover new truths, and help us not only to find but also to choose ourselves.

This kind of pretending is illustrated by my discovery on the river: "When I act brave, I'm better able to be brave." Much the same thing happened to Tim, who acted more mature and self-focused than he actually felt in order to reach for his own emotional competence. A bold act of pretending can help expand what is real and true about ourselves and our relationships.

MUST YOU ALWAYS HAVE IT ALL TOGETHER?

Some of us need to practice sharing more vulnerability, not less—especially if we're entrenched overfunctioners. Overfunctioning is not simply an overzealous wish to be helpful but a patterned way of managing anxiety that grows out of our experience in our first family. For example, an eldest daughter may have tried to keep her chaotic family afloat, learning along the way that it's too painful and disappointing to reveal her own needs and expect them to be met.

If we overfunction as adults, we tend to know what's best not only for ourselves but also for others. We have difficulty staying out of other people's problems and allowing others to struggle toward their own solutions. We strongly resist voicing our vulnerability with an underfunctioning family member, and we may be convinced that he or she has nothing to offer us anyway.

Making even a small dent in our overfunctioning ways can foster mutually empowering connections, and it can lead to a more accurate picture of our self and the other person. Experimenting with conversations that initially feel unlike our "real self" can lead to real change. In giving voice to our limits and vulnerability, we can wake up those aspects of our self that have been suppressed or shamed into silence. The world of family relationships is a great place to start.

JANET AND HER SISTER, BELLE

Janet initially came to see me for work problems, but she was also having a hard time with her younger sister, Belle, who was still reeling from the trauma of her divorce two years earlier. "My sister always has some massive area of worry," Janet told me. "Her problems fill up the room."

I learned that their relationship was rigidly polarized, with Belle in the role of the emotional basket case and Janet in the role of the competent firstborn daughter. When I first encouraged Janet to talk to Belle about her own problems, she saw no point in doing so, because Belle always shifted the conversation immediately back to herself. Janet tried to avoid talking to Belle, because of her irritating self-absorption and expressed helplessness.

When Belle first divorced, it made sense for Janet to put aside her own needs to be available for Belle. But Belle's crisis had become chronic, as had the imbalance in their relationship. Indeed, the pattern Janet described (Belle underfunctioned, Janet overfunctioned) had deep roots in old family roles. Janet's irritation with Belle's " 'poor me' routine," as she called it, was also a clear signal that she couldn't continue this way and still stay well connected to her sister.

"Let Tell You about *My* Bad Day"

I encouraged Janet to be persistent in sharing her own difficulties with Belle, even if she started with a small step like telling Belle she'd had a bad day. Janet's initial attitude could be expressed, "Belle obviously isn't able to be helpful to herself, so what could she ever offer me?" and "It's not worth the effort." But it was worth Janet's effort. It's not fair to allow another person to dominate the conversation and then to blame her for it.

Instead, Janet needed to go the extra mile—to call Belle on the phone, for example, and say, "I had a terrible time at work today. I'm so glad you're home, because I need to tell you about it." When Belle shifted the focus to herself, Janet could say, "Belle, I know you're struggling, but to be honest, I just can't listen very well to your problems right now. I'm upset about what happened at work, and I want to get your opinion." When Belle went on longer about her ex-husband than Janet could comfortably tolerate, Janet could aim to be light and funny in pointing out her limits: "Belle, if you mention that ex-husband of yours one more time today, I think I'm going to hide under a rock! That guy has been occupying far too much space in our conversation!" She could tell Belle when she was too tired to talk or pay attention.

Substantive change doesn't occur in one hit-and-run conversation. Family patterns change slowly, sometimes at a glacial pace. It's the direction we're moving that matters, not the speed of our travel. Nor is the goal simply to get results, which can never be guaranteed. Rather, Janet will be on more solid ground in all her relationships if she can practice voicing her vulnerability and limits with her sister over time.

We all resist change even as we seek it, and Janet was no exception. At first she saw no point in practicing new conversations with Belle or pretending that she needed her. Janet was convinced she didn't need anybody in her family, and certainly not Belle. But when Janet looked at her life less defensively, she realized that she did need her sister. She also recognized that her inability to show weakness or vulnerability in her other relationships created a certain distance. Janet actually felt pretty isolated in all her relationships, except when she was helping, giving, and being of service. She was pretty tired of doing so much for other people and often felt that she didn't get much in return, although she didn't see her own difficulty letting others in.

Preparing a Script

When Janet did consider voicing her vulnerability and limits with Belle, she became so anxious she couldn't think straight or find the right words. So in therapy, I helped her plan how she might address her concerns directly when Belle again changed the subject back to herself. For example, Janet might say:

> Belle, when I try to talk to you about my problems, I get the feeling that you're not really there for me. I know you're still in a great deal of pain about the divorce and your kids, and I do want to be available to you. But you're my sister, and I need you to be here for me too.

Or Janet could speak warmly and directly to the lack of balance in their relationship by saying something like the following:

> Belle, I get the feeling that you think my problems aren't important because they are so much smaller than yours. Your problems are bigger than mine right now, but my problems are important to me, and I need you to listen when I'm upset.

In role-playing such conversations in our therapy session, Janet joked that she might need CPR after uttering the first sentence. She felt wildly anxious at the very thought of sharing her vulnerability with her sister and reaching for Belle's competence to respond supportively. I suggested that she prepare a "script," which gave Janet a plan for controlling her anxiety. The script also helped Janet to avoid sounding edgy or irritated, so Belle was able to hear her comments not as a criticism but rather as a request from a sister who loves her and values her perspective.

In the process of reading from her script (a staged and superficial idea, on the face of it) and pretending that her sister had something to offer her, Janet came in touch with her authentic buried needs. She recognized how incredibly hard it was for her to persist in voicing her wants, especially in a manner that truly invited the other person to attend to her.

Get Off Your High Horse

We are not doing any favors to a helpless-appearing family member or friend by failing to share our own problems and complaints. When we only listen and try to help—and we don't share our own limitations, vulnerabilities, and worries (we all have them)—we act as if that other family member has nothing to offer us and isn't capable of showing some caring. We may be firmly convinced that this is so, but we're denying the other person the opportunity to rise to the occasion and feel useful.

To reach out for the other person's competence—even when it isn't readily apparent—is an act of respect. As Goethe wrote (before gender-inclusive language): "If you treat man as he appears to be, you make him worse than he is. But if you treat man as if he already were what he potentially could be, you make him what he should be." We can never know the totality or full potential of other people (or what they "should be," for that matter), but who they are with us has something to do with who we are with them. Through our conversations, we unwittingly enlarge or diminish the potentials and possibilities of everyone around us.

My clients who are depressed may say something like, "I'm not going to tell my sister (or, mother, husband, friend) I'm feeling depressed, because she's *really* depressed." We may think we're being considerate by not "burdening" a person who is already burdened, but the opposite is true. The least helpful thing we can do is to keep focusing on their problems and trying to be helpful to them. Instead, it would be more helpful for us to begin to share our own problems, limitations, and needs.

No one benefits from a polarized relationship where we listen, help, and offer advice, then say, "I'm fine," in response to the question, "How are you?" We diminish people when we don't allow them to help us, or when we act like we don't need anything from them and they have nothing to offer us. We also diminish them when we allow them to go on and on, even after we've exceeded our capacity to pay attention.

More to the point, our own self-regard ultimately suffers when we're unable to present both our competence and our vulnerabilities to the key people in our lives. Janet did ultimately move toward sharing a more balanced self-presentation with her sister, but not "for Belle," and not even for the sake of eventually getting the response she wanted. The truth is, we may never get the response we want from the other person, no matter how hard we try. But as Janet experimented with sharing the softer, more vulnerable parts of herself with her sister, she learned to share a balance of her strength and vulnerability in other relationships.

MY BRILLIANT SISTER

It takes enormous courage to pretend, experiment, or act "as if" in the service of sharing a self that is more whole. When we can present a more accurate, balanced experience of our self to our family members, we can begin to perceive other people with greater accuracy, including an intimate partner. Let's see how that works.

My sister Susan, a typical firstborn, overfunctioned with a vengeance when we were growing up. She was my mother's helper, my father's pride and joy, and a model child. Her role in the family was that of the perfect child and the brilliant star, shining larger than life in my father's eyes. For many years, I underfunctioned with an equal vengeance. I was "the mess," the difficult child, the one with the problems. Since family roles can be rigidly enforced, my status as the intellectual underdog persisted, no matter how well I did or how hard Susan struggled. More than once my father would announce to a disinterested acquaintance, "Harriet is bright, but my other daugher, Susan, is brilliant." And also, "I don't think anyone is as brilliant as Susan." At times, his unabashed labeling of Susan as the most perfect person to inhabit the planet reached truly bizarre proportions, such as when he told me how jealous *all* the other parents became on just seeing Susan in her stroller—how they just wanted to swap their own baby or toddler for Susan.

Being an idealized child carries a high price tag. It interferes with healthy self-esteem, which requires us to have an objective view of our strengths and limitations. Being devalued also interferes with objective self-knowledge, but we're more likely to protest or resist labels that diminish us. Idealization is seductive. But as someone wisely noted, a pedestal, like a prison, is a small space in which to navigate.

Susan was indeed gifted and creative, and she possessed a towering competence. Whenever the extended family got together, her role was to regale folks with her interesting stories and adventures. She was an avid reader, had an encyclopedic mind, and spoke with detailed assuredness on all subjects—even those she actually knew nothing about. But she didn't ever share her vulnerability or act as if she needed anything. It was terribly hard for her to say, "I don't know." Our sibling relationship was totally out of balance, strained by our polarized roles in the family and by my own sense that Susan knew everything while I had nothing to offer her.

Of course, all was not perfect in Susan's life. One area where her superior intelligence failed her was with men. She seemed unable to see men accurately or to use her intelligence to judge their character and intentions. I was aware of her problem, but Susan had never addressed it openly in the family, and I hadn't taken the initiative to bring it up.

A significant turning point came when Susan visited me in Topeka one Thanksgiving. I was scheduled to meet with a family systems therapist and, half-jokingly, I offered her my session as a gift. To my surprise, Susan accepted my offer and asked me to come with her. I listened, wide-eyed, as Susan plunged right in, detailing her history of intense starts, poor judgment, and lack of confidence in her ability to evaluate men. She said she felt like a leaf tossed about by the wind when it came to love relationships. I was deeply moved by her openness, and by her resolve to get a grip on her problem. I was also impressed that when the therapist gave her an extremely difficult assignment, she ran with it.

Beginning the "De-Princessing" Process

The therapist suggested that Susan "de-princess" herself with our father, Archie. Part of the theory was this: If Susan could move in the direction of sharing with Archie a more balanced and objective picture of both her competence and limitations (and also get to know him better as a real person), she would be better able to develop a more accurate view of the men she was dating and show them a more accurate view of herself. If she continued in the role of "Daddy's perfect princess" (a role so entrenched that it felt real), the cost would be less objectivity and balance in her relational world.

Susan's first de-princessing task was to write Archie a letter sharing her difficulties with men. She explained a bit about the problem and said that she didn't understand why this was a struggle for her, but she wanted to tell him in case he had any ideas or suggestions. Letters had always been my father's most expressive medium, so writing rather than calling made sense. The therapist suggested a separate letter to my mother, also eliciting her perspective.

The proposed exchange violated two family rules. First, Susan wasn't supposed to acknowledge her own personal limitations or problems. Second, opening up to our father—as if he actually had something to offer on the emotional front—contradicted our tendency to avoid bringing anything real or substantive to his attention. Archie was entrenched in the role of the distant outsider in the emotional life of our clan, while Susan and I were bound together with our mother, Rose, by unswerving loyalty. Our unspoken family rule was that Rose was the only one we were supposed to talk to about anything personal.

Surprisingly—or not surprisingly—my father wrote back a thoughtful response to Susan's letter. Several conversations through letters followed, in which Susan voiced more of who she really was in love and work, and Archie penned evocative letters in response. Within a year or so, Susan met a wonderful guy whom she married as she turned fifty.

Now, please don't assume that if you just overcome your old dysfunctional habits and give voice to your full authentic self to your difficult father, you'll have no problem finding Mr. Right. Yes, finding an intimate partner is partly a matter of emotional readiness—but it's also a matter of opportunity, hard effort (you can't wait to be discovered) and, yes, plain old-fashioned luck. I should also mention that Susan began to maximize her opportunities to socialize and reminded her friends that she wanted their help meeting suitable men. One of her lifelong friends introduced Susan to the man who became her husband.

The point of this story is not that Susan married. Although she wanted an intimate relationship, she had an abundance of friends and was quite satisfied living on her own. What mattered was that over time, she worked on several of her relationships to allow herself to be known and seen in a real way. This, in turn, helped her to know and see others more accurately. She achieved this by broadening the conversation about herself, by being *real* instead of *ideal*.

The changes Susan made produced a ripple of positive effects in all the relationships in our family. Today Susan and I are closer than ever before, free from the overfunctioning, underfunctioning polarity that once defined our relationship. My father, for his part, did not show an enormous flexibility for change. But when Susan opened up to him as if he were capable of more insight and connectedness than he showed, he did raise his level of functioning a bit. And small changes can go a long way.

Acting "As If"

Tim was in an underfunctioning position in his situation, while Janet and Susan were card-carrying overfunctioners. But all three faced the challenge of changing their part in polarized relationships. Paradoxically, it was their willingness to experiment and pretend that allowed each of them to construct both an "I" and a "we" that was more complex, richer, and more accurate. When parts of the self have long been silenced—or relationships are out of whack—it can be honorable and life-changing to act "as if." This is especially so when our goal is to enhance rather than to diminish the self and others and to test out what's possible in a relationship.

CHAPTER 6

Putting Our Parents
in the Hot Seat

Many years back, I was riding in the car with my father while visiting my parents in Phoenix. When he stopped at a red light, a woman about my age crossed the street in front of us. She had extremely frizzy brown hair, the sort that misbehaves, has a life of its own, and doesn't flop back down into place when disturbed.

"*Look* at that hair!" my father said, with utter dismay. He shook his head in grave disapproval. "Just look at that *hair!*" he said again. "What a *mess!*"

Of relevance to this exchange is the fact that my own hair was identical to that of the woman he was exclaiming about.

"Daddy," I said, tapping his shoulder so he'd turn in my direction. "Look!" I grabbed a clump of my hair in my fist and held it out to him. "I have the exact same hair."

This particular interaction occurred during one of my mature moments, so I was entirely calm. I didn't act wounded or angry. I wasn't sarcastic. Nor did I mumble or toss out my comment into the air, which would have made it easier for him to ignore me. I was present, right there, and reaching for a connection with him. My tone was warm and curious, but my manner didn't let him off the hook. I was making an inquiry—"Well, Dad, this is interesting. What do you make of it?"

My father cleared his throat. "Well, I was talking about her," he said.

"But don't you think she and I have the same hair? Just look."

"It doesn't matter," my father said matter-of-factly, again inviting the conversation to come to an end.

"Well. Daddy, I want you to know that your comment hurt my feelings. Like I said, I have the same hair."

My father said nothing. The light changed, and we moved on to other subjects.

I recently shared this snippet of conversation with my friend Jennie in response to a story of her own. She was furious at her mother, who had referred to Jennie's friend as a "fatso." Jennie asked her mother to keep her rude opinions to herself. Her mother replied, "Well, I'm a fatso, and so are you and your sister. All the women in our family are fatsos."

Jennie felt devastated at this jab at her own weight, a sensitive issue for her. She took the conversation no further. When I told her about my conversation with my dad, she couldn't relate to how I handled myself.

"Why in the world would you make yourself so vulnerable?" she asked. "I would never put myself in such a position."

Actually, my father was the one who had been put on the spot. He was the one who was forced to sit in the hot seat for a minute right there in the car. It's not that I thought my words would change him. Rather, I spoke because I believed it was the honorable thing to do, and because I wasn't going to protect him by ignoring his comment and allowing him to avoid considering how his words affected me. At the same time, I was focused on what I wanted to say *for myself,* not on getting a particular response.

When someone hurts our feelings or behaves badly, we typically respond with anger or silence. It's normal to react this way, but when we do, we may be letting the other person off the hook. We feel as if we're protecting ourselves, but we may actually be more concerned about how other people will manage themselves in a difficult conversation and how uncomfortable they might become. We may be nervous about gently putting the other person on the spot. What if Jennie had said to her mother: "You know, Mom, the comment you made about the women in our family being fatsos hurt my feelings. I'm sure that wasn't your intention, but my weight is a sensitive issue for me. I sometimes make negative comments about being overweight, but it feels different to hear my own mother call me a fatso. I wonder if you're aware of this?" What if Jennie then stayed calm and left space for her mother's response? Who is Jennie actually protecting when she concludes that there's "no point" in speaking up?

Obviously, we don't have to address every insult and injustice that comes our way. It can be an act of maturity to simply let things go. With family members, however, there's a lot of mileage in finding the courage to speak clearly. Strengthening our voice with key family members will affect every relationship we're in because other relationships get overloaded when we can't talk to family members about what matters. Everything is interconnected. As the naturalist John Muir put it, "When we try to pick out anything by itself, we find it hitched to everything else in the Universe."

If you increase your level of functioning in one key family relationship, every other relationship will change. The degree to which you can be clear with your first family about who you are, what you believe, and where you stand on important issues will strongly influence the quality of "voice" that you bring to other relationships.

MY DAD MAKES CRUDE COMMENTS

Anna sought me out for help in dealing with her difficult father. Following her parents' bitter divorce fifteen years earlier, he had disappeared but had now resurfaced after her mother's recent death. He genuinely appeared to want a relationship with Anna, and in many ways he acted generous and kind. But when he and Anna were out in public, he often made crude sexual comments. He'd stare at women's breasts and say, "Wow! Look at those big ones!" or "I'd like to see her without her sweater!" Anna found it intolerable that her long-lost father behaved this way in her presence. She was furious at him, but she also worried that he'd disappear again if she confronted him.

From my perspective, his provocative comments were probably an expression of his high level of anxiety. Both he and Anna were dealing with the death of Anna's mother, which would inevitably stir up old family history, and on top of that was the enormous challenge of trying to reestablish their own relationship. These are not small emotional events. Indeed, I told Anna she should expect to deal with some difficult fallout for at least a couple of years.

When Anna first came to see me, she had written her father a long letter confronting him about his behavior. The unsent letter, which she brought to therapy and read to me, covered—among other small topics—sexism, feminism, and his objectification and devaluation of women. These were all important issues, but even Anna could predict that the letter might only make communication more difficult. Letters to family members can be a helpful way to widen the path for truth-telling when we're able to open up a loaded issue in a brief and nonblaming fashion. But long, confrontational letters almost always shut down the lines of communication and evoke defensiveness rather than fostering understanding or mutual empathy.

I suggested to Anna that she not mail the letter. It would only fuel her dad's anxiety and increase the likelihood that he would continue to behave obnoxiously or even take off again. When we criticize people or lecture them, we actually invite them *not* to pay attention to what we are saying. It was impressive that Anna and her father were able to even be together at all after his long absence, especially in the aftermath of Anna's recent loss. I encouraged Anna to be patient with her father—and with herself—as she worked to stay connected and to formulate a way to respond to his inappropriate comments.

At first, Anna saw only two options—to mail the letter or to ignore her father's comments. But staying silent when her dad's behavior was inappropriate and upsetting was unfair to Anna, to her father, and to their relationship. Because the situation was incredibly tense, I helped Anna formulate an alternate plan for what she would say to her father. She wrote out what she wanted to tell him and then rehearsed it with a friend.

It may strike you that such strategic planning is the opposite of honesty, but as I said earlier, that's not the case. Sometimes we can wing it, but when the stakes are high, we need to be thoughtful and well prepared. Of course, if Anna's only goal was to voice her anger at her dad, and to give him the full brunt of her emotions, she could simply do that. If, however, her goal was to be heard, and to give the relationship with her father the best chance of succeeding, then she needed to plan ahead. It is possible to speak honestly and also to proceed with care to protect a relationship that is important to us.

Short and Brave Exchanges

Here's what Anna did. The next time her father said something inappropriate, she said calmly but firmly, "Dad, when you make comments like that, I feel very uncomfortable. Please don't talk that way when you're with me."

If Anna's dad happened to be extremely open and flexible, one comment might produce the desired results. But change rarely happens that way. Predictably, he reacted defensively. He said things like, "You're just being oversensitive. Don't tell me how to behave!" Anna's automatic response—good feminist that she was—would have been to get angry in return, and to confront his sexism and objectification of women, as she had done in the letter she scrapped. But instead she spoke about herself—his impact on *her*. She said, "Dad, I need to let you know how anxious and uncomfortable I feel when you stare at women or make comments about their bodies." It took enormous courage for her to say these words as she tried to maintain her connection with him. Therapy helped her to not take her dad's comments personally, but rather to see them as expressions of his high-level anxiety as he moved back into the anxious emotional field of the family he had fled from.

When addressing anxious issues, it's important to have a process view of change. Substantive change in families doesn't occur with hit-and-run confrontations. Anna needed to give her dad some time to think about what she was telling him, and her dad was likely going to test whether Anna "really meant it" and whether she would hold her ground with him. He initially responded with extreme defensiveness, and he continued to make inappropriate remarks in Anna's presence, although now he muttered them almost under his breath.

Anna decided that the next step was to drop him a note that restated her position.

Dad, I had a really nice time with you the other day. It means a lot to me that we're back in touch after all these years. I've lost Mom. I don't want to lose you again. You're very important to me. But I want to tell you again that it's difficult to be with you when you continue not to respect my feelings. I'm your daughter—not your buddy. I don't think the stares and comments about women's bodies are appropriate in front of me. It certainly is my hope that you'll think about what I'm saying and consider my feelings.

Love, Anna

Although her father didn't mention the note, his sexual comments dropped to near zero. Anna made the interesting observation that her dad occasionally reverted back to his old behavior when they had more frequent contact or when one of them pushed for more of a relationship. I suggested to Anna that she monitor how much time she spent with her father. Sometimes, when family members reconnect after a long absence, one or both parties try to accomplish too much too fast. Reconnecting with a parent after a long period of time is best accomplished slowly. Even if Anna or her dad wanted to establish a close relationship, it was important for them to understand that healing after a lengthy cutoff is a slow process that usually moves forward in fits and starts.

I wasn't suggesting that Anna retreat into cold distance, but rather that she take smaller steps in getting to know her father, such as experimenting with less contact or keeping their conversations light and bantering. Her father's inappropriate comments almost always occurred when they were walking together. So if her father didn't change, Anna could have said, "Dad, I'd love to meet you for dinner at the restaurant. But I don't have time for a walk afterward, because I have work to do."

If Anna were dealing with a less key player, she could have handled the crude comments by telling the person off or by refusing to hang out with him. But this was her father. It helped Anna to appreciate the high level of stress she was under at this time and the enormity of what she was trying to accomplish. She was being called on to cope not only with the loss of her mother but also with her dad's move back into her life. Either of these events by itself would be enough to make a perfectly normal person freak out.

The most difficult conversations Anna initiated with her dad occurred about a year and a half later. Anna asked her father to help her understand his disappearance from her life, and what had kept him from staying in touch with her when Anna's mother was still alive. These were painful and important exchanges that never would have taken place if Anna had made the move to "kick the bastard out of my life"—her initial response to her father's crudeness.

OPENING A CLOSED MIND

We may choose to confront an insensitive family member directly, forcefully, and immediately. Sometimes turning up the volume and intensity does the trick, and it can definitely provide a momentary sense of relief. There is no one "right" or "best" way to speak, or to change the hearts and minds of others. But if our old ways of speaking or staying silent are bringing us pain, it won't help to do more of the same. It also matters if our goal is to vent our immediate feelings, or to widen the possibility for conversation and truth-telling over the long haul.

Doing the "Two-Step" around a Very Hot Issue

Joyce sought my help at a point when she was furious with her mother. Her sister's wedding was coming up, and Joyce was planning to attend with Melody, her partner of thirteen years. Her mother phoned and said, in a snippy tone of voice, "Remember, this is your sister's wedding, so you don't need to bring attention to your relationship with Melody." Joyce said (even more snippily), "Oh, thanks a lot, Mom. Maybe I'll have Melody put a paper sack over her head to spare you from any embarrassment." Joyce then quickly filled the silence by saying, "I'm out of here," and hung up the phone.

What can we make of this mother's hurtful comment? Weddings are predictably anxious times in the life cycle of families, so it's no surprise that Joyce's mother was looking for a place to focus her intensity. In all fairness, she didn't invent homophobia on her own but rather had learned the prejudices our culture teaches. Plus, she was further along than many parents, in that she hadn't cut Joyce off and she openly acknowledged Melody as her daughter's lifelong partner and not just her best friend. As for Joyce, of course, she was angry that her sister's wedding was being celebrated in grand style, while she was asked to treat her most important relationship like an embarrassing secret. How could Joyce not be upset that her mother wasn't at the level of acceptance Joyce would like?

Joyce felt she had stood up to her mother by responding to her snippiness with sarcasm. Actually, she protected her mother by responding in kind and by not taking the conversation a step further. When her mother said, "Remember, this is your sister's wedding, so you don't need to bring attention to your relationship with Melody," Joyce might have taken some deep breaths and calmly asked questions that would invite thinking rather than reactivity. Like, "Mom, I'm not sure what you mean by 'bringing attention' to my relationship. Can you tell me a little more about what your concern is?"

Other questions might follow. Would there be family members or friends at the wedding who didn't know that Joyce is a lesbian or who didn't approve? Who among the guests did her mother anticipate would have the strongest negative reaction? Who in the family has had the most difficult time accepting Joyce's relationship with Melody? Was her mother aware that Joyce felt hurt by her comment, even if that wasn't her mother's intention? After hearing her mother's point of view, Joyce might simply have said, "Mom, I know how much stress you're under with the wedding coming up. And I know how upset your sister gets when she sees Melody and I holding hands. Of course you want the wedding to go perfectly. But

I need to tell you that your comment about not bringing attention to my relationship with Melody hurt my feelings."

Joyce did not need to take the conversation further at that moment. If we don't feel emotionally ready or prepared, we can return to a subject or conversation later. Joyce's challenge is to ask something—or to share something about her partnership with Melody—without needing her mother to approve or respond in a particular way.

Over time, we can expand the conversation. Sensitive issues are processed slowly. Joyce described herself as an "out-there, in-your-face person." But she'd been out to her parents for twenty years and had never asked, "Mom, what's the hardest part for you about my being gay? How has it been for you?" Instead of using her creativity to generate questions, she had created reactivity and distance. This is not to criticize Joyce, who was simply responding as we all do when we feel anxious and wronged—resorting to what's automatic or familiar, and not taking the risk of moving the conversation to a deeper and more authentic level. It may seem as if snapping back at someone gets the point across. But nonreactive, engaged conversation is far more likely to get at the root of hurt feelings and change them.

Practice! Practice!

How do we learn to ask questions, define our differences, and stay relatively calm and clear when we don't get the response we want? The only way to learn to speak is to speak. You may know the old joke about the guy who approaches a street-corner musician in New York City and asks, "Excuse me, sir, how do I get to Carnegie Hall?" The man replies, "Practice!"

Practicing is necessary to accomplish anything worthwhile, and speaking up in a difficult relationship is no exception. We can start with small steps or easy issues. Then we can practice asking clear questions about the very subjects we most want to ignore. Next, we can define where we stand and address the differences.

When a particular topic (or person) is especially difficult, remember the "two-step." Try to think in terms of having at least two conversations, or a series of conversations that fall into two categories.

In the first conversation, we only listen, ask questions, and try to learn more. For example, "Mom, what's the hardest thing for you about my being gay?" "How do you imagine Grandma would react if you had been gay?" "How did she respond to Uncle Charlie when he left the priesthood and married a Japanese-American wife?"

In this way we let the other person know that we are genuinely interested in learning more about their perspective. Listening is an essential part of having voice. As we enlarge the context around a problem ("Mom, it sounds to me like Grandma wasn't very tolerant of people who were different"), we take the pressure off the hot spots and begin to better understand where the other person is coming from.

We feel calmer as we come to understand that the other person's insensitive response is fueled by anxiety and history—not lack of love. It's an act of maturity to *not* take things personally and to understand that the other person's response may have more to do with them than with us. Our thinking and voice will be clearer to the extent we can view our parents' negativity simply as *information* about their way of managing anxiety. In a subsequent conversation, we can share our perspective and define our differences. For example:

> Mom, I was thinking about our conversation last week. You and I see my being gay quite differently. As I understand it, you think that it's a problem I was born with, that I can't help the way I am, but that you love me anyway. I have a

very different view. When I was first in touch with my feelings toward women, I felt scared and I wondered what was wrong with me. But now, my relationship with Melody is the best thing that ever happened to me. I feel fortunate to be who I am. If I could push a button and magically turn heterosexual, I would never do it. What do you think about this difference between us?

Speaking to the differences is not the same as trying to convince or change the other person. It doesn't imply that the other person is wrong and that truth is on our side, although we may be convinced that's so. Instead, it requires us to clarify and refine our differences with as much respect for the other person's different perspective as we can muster. This respect, and our willingness to listen, can be contagious.

But that's not why we do it. When speaking to any hot issue with a family member, we should stay focused on what we want to say about ourselves, rather than on eliciting a particular response from the other person. If we're *needing* (as opposed to hoping for) a particular response from the other person, that's a good indication we're not yet ready to broach a difficult conversation.

Like many things, this two-step process sounds simple in theory:

First, ask questions and listen.

Second, speak to the differences.

But these two steps are enormously difficult to put into action. When you're dealing with a high-twitch subject, your brain will turn to mush. You won't have a clue what questions to ask. The concept of "speaking to the differences" will totally elude you. If you're drowning in emotions, you won't draw on your creativity or even your common sense. You'll get critical, defensive, or just plain mad. When this happens—or ideally before it does—you'll need to think and plan. Find a clear-thinking friend to help you walk through the process, because it's almost impossible to apply your best thinking to your *own* family.

As Jon Kabat-Zinn reminds us, the human mind is like the surface of the ocean that gets whipped about by bad weather. We all get reactive, but underneath the waves there is a deeper calm if we can only reach for it.

WHEN EMOTIONS SPEAK LOUDER THAN WORDS

I don't mean to imply that we should never confront a family member with the full force of our emotionality. Perhaps you can think of instances in your own family where doing so is powerful and helpful—or just plain unavoidable. It's ridiculous to suggest that we are always going to be able to speak in calm "I-language," or even that this should always be our goal.

I once worked with a client, Francine, who was a very overcontrolled, in-charge first born. She would repeatedly tell her alcoholic sister, "I know I can't do anything to fix your problem, but I want you to know how scared I am of losing you. I feel very sad when I think about you not being around for as long as possible." Francine happened to be a therapist herself, and indeed she sounded like one whenever she addressed her sister's drinking.

One night Francine lost her temper and started screaming at her sister, "What's wrong with you? How can you throw your life away like this? How can you do this to me! I can't stand it anymore! I feel like I'm going crazy!" Then Francine crumpled on the couch and sobbed. Her sister just looked at her and walked out the door. Francine thought she had blown it. Instead, it turned out to be a major turning point in their relationship. Her sister had long resented Francine's calm, therapeutic tone, which felt like a subtle form of arrogance and one-upmanship.

★   ★   ★

We have so many different voices and places within that we can speak from. The challenge is to know our full range so we don't get stuck in a narrow, habitual form of self-expression that isn't serving us. We obviously speak differently to different people and to the same person at different times. We don't do this because we're wishy-washy chameleons. Rather, different people evoke different parts of ourselves. Also, we know that different people have different receptors for what they can hear and respond to. Some relationships tolerate a much wider range of expression than others—whether what is being expressed is silliness, judgment, adoration, or anger. We all need connections with people who make our voice larger, not smaller.

Family relationships tend to be intense. They may appear calm, but that's often because the intensity is managed by distance. As a rule, the higher the intensity, the more productive it can be to slow down, to ask questions and listen first, and then to calmly speak to the differences. In this way, we invite the other person to feel accountable rather than defensive.

In certain circumstances, however, our willingness to show our emotions and vulnerability is the most powerful message of love we can convey to another family member. This was certainly so for Francine. There is more than one way to put a family member on the hot seat, and more than one way to use our voice with love.

"DON'T YOU *EVER* DO THIS AGAIN!"

Let me share a conversation with my mother where I threw all caution to the wind. I didn't calmly ask questions and define my differences, but instead I jumped up and down and told her she darn well better change, or I'd be one unhappy puppy.

It's ancient history, yet I recall the following exchange as if it happened yesterday. On a Saturday morning during the summer of 1986, I was working in my office. I was catching up on paperwork, having fallen behind after the birth of my first son. I thought I was alone in the building when my husband, Steve, startled me by his unexpected arrival. I knew right away that something was wrong. "I have bad news about your mother," he said. "She had a breast removed yesterday. She has breast cancer. Or she had breast cancer." He didn't know which tense to use—a common conundrum in talking about this particular disease.

I pulled away from the arms he held out to embrace me and fought back nausea. How could this be? I had just chatted with her on the phone several days earlier, and nothing seemed out of the ordinary. We'd just had a "How are you, how's the weather?" sort of conversation. Now I learned that she had been heading into major surgery with a serious diagnosis. As I mentioned, my mother had had an earlier cancer surgery when I was twelve, and no one had said a word then—before or after. But that was a very long time ago, and circumstances were very different now.

I drove home and immediately called my mother in the hospital. Maybe I took a minute or two to ask how she was feeling and to get the facts. Then I gave it to her.

"Don't you *ever* do this again!" I ordered with unbridled passion. "I love you *so much.* How could you *do* this? How could you *not* tell me what was going on?" My speech was filled with italics. I was anything but low-key.

"Well, I didn't want to worry you," my mother said thoughtfully. "I decided to wait until after the surgery was over before I told you and Susan."

"You didn't want to *worry* us!" I was incredulous. "Mommy, this is what family is *for.* We *worry* about each other. It's my *right* to worry about you! You're my *mother!*"

My mother chuckled. I knew that she felt the love that was fueling my passionate insistence for this deception never to happen again.

"What if you had died in surgery?" I continued. "What if I didn't even get a chance tell you I love you!"

"I know you love me," my mother replied warmly.

"What if I wanted to *pray* for you?" I argued.

"You don't pray, Harriet," my mother reminded me. I could picture her smiling at the other end of the phone.

"Well, maybe I *would* have prayed for you *then*," I insisted. "How do you *know* whether I would have prayed or not? Maybe I would have asked someone *else* to pray for you."

My mother was silent. I recognized that the conversation was getting a bit off track.

"Look, Mommy," I said, "promise me that you will NEVER do this again! Nothing is worse for me than what just happened. Nothing! I can *deal* with worrying about you. I can deal with *anything* that happens to you. But I *can't* deal with worrying that something serious *might* be going on and you're *not* telling me. THAT'S what I can't deal with."

"Okay," my mother said. "I'll tell you."

"If you *don't* tell me, I'm going to worry ALL THE TIME, because I won't know when to really worry." I could not make my point strongly enough. "Promise me that you'll *never* do this again!"

"Okay," my mother said again.

"*Promise?*"

"I promise."

That was that. But just in case my mother hadn't taken me seriously enough, I called my sister, Susan, and asked her to join forces with me. She hopped right on the bandwagon and called Rose too, since she felt exactly the same way I did about this matter. Our mother, from that day forward, has certainly kept her promise.

A friend who knows this story was taken aback by my behavior. "If Rose was facing surgery, she needed to do whatever was right for her. You have to admit that the decision to share health information is a personal choice. Why would you try to make this decision *for* her?"

Of course I couldn't make the decision for Rose. No one has that sort of power. If Rose ultimately decided not to share health information until after the fact, I would have come to terms with that. But Susan and I were certainly going to disabuse her of the notion that her silence would offer *us* some kind of protection. We were crystal clear that we didn't want to be the sort of family that "protected" each other by not sharing health information. Yup, we jumped up and down and stomped all over the place on this particular issue.

"Perhaps your timing was a bit off?" My colleague's gentle question referred to the fact that I didn't wait several days—or even five minutes—before jumping in. What he really meant was, "Your mother just had major surgery, for God's sake! Couldn't you wait until after she got home from the hospital before bulldozing her?" Perhaps he was modeling the good communication skills that I had failed to demonstrate in confronting my mother.

I thought about his concern. If Rose had become upset or was just not up to the conversation, I would have returned to the topic in a new way at a different time. I was listening to her, and I could hear right away that she was listening to me. I knew that she'd have a positive response because my words were fueled more by love than by anger.

"I know my mother," I said to him, and that's really the point.

We all need to rely on a combination of intuition and thinking to decide how and when to put a parent—or anyone we love—in the hot seat. It's nice when we can approach a difficult conversation with the expectation of being heard and considered, as I did with Rose—but obviously we can't count on that. The other person may be totally unable to hear us, and we may still decide to speak with clarity and intention, to take the conversation another round, for the sake of honor and our own personal integrity.

Obviously, people are most likely to hear us if we are letting them know that we love them and that they are important to us. That we may not get the desired response is painful but beside the point. The conversation I had with my father in the car about my hair wasn't intended to get to him, or to move him, or even to evoke an apology or behavioral change. I knew from past experience not to expect this, just as I knew not to expect my hair to behave. But I wasn't going to protect him either. I simply needed to hear the sound of my own voice speaking to my father without backing down.

CHAPTER 7

Love Can Make You Stupid

A friend confessed to me recently that she wasted a full hour before her date arrived with the inane activity of arranging and rearranging the magazines on her coffee table to make just the right impression. Then she camped out by the front window awaiting his approach, so that she could dash to the CD player to ensure that a particular song would be playing as she opened the door. Later they went to a movie that she found so offensive that she could hardly stay in her seat, but when *he* left the theater raving about it, she didn't share an honest response.

We can act like this in the beginning, but we can't sustain it for too long if the relationship is to move forward. Over time, we must move in the direction of greater authenticity, especially when we feel the relationship is significant and we want it to endure. The more intimate the relationship, the greater both the possibility and the longing to share ourselves—and the bigger the emotional consequences of not telling, of not being real, of not bringing our full self and true voice into it.

After all, if a potential partner doesn't want to stick around after a well-meaning revelation that we had an abortion, a mastectomy, two previous marriages, or a recent Nobel Prize nomination, we're better off without that person. We may also be better off without him if he doesn't especially like hearing our exuberance, expansiveness, and ambitions, or if she shuts down when we voice our insecurities, fears, or a painful story from the past. Likewise, the other person has a right to know us accurately, to consider the relationship and make plans for the future based on facts, not fantasies or projections. Intimacy—and our judgment about the relationship—suffers in the shadow of silence and pretending, which does not allow us to know the other person or to be fully known.

We've seen how some kinds of pretending can be brave and enhancing, but that's not the sort of pretending I was raised on. When I was growing up in the 1950s, before modern feminism, we were taught to "play dumb," let the man win, pretend he's boss, and listen wide-eyed to *his* ideas, no matter how boring, gracefully adding a footnote from time to time. Anything to get and keep a man. Grown women behaved like female impersonators, to use Gloria Steinem's wonderful phrase.

Today we're bombarded with messages everywhere encouraging us to speak authentically and truly. Columnist Ellen Goodman has a friend who wisely encourages each of her three little girls to "Speak up, speak up, speak up," with the frank explanation, "the only person you'll scare off is your future ex-husband!" That's a fairly radical lesson—that having a man shouldn't occur at the expense of having a self. But despite lots of good advice out there, love and romance do not tend to foster the expression of a clear and strong voice. As my friend the cartoonist Jennifer Berman notes, pairing up is more likely to reduce the "judgment lobe" of our brain to the size of a pinto bean.

FALLING IN LOVE TELLS US NOTHING

Falling in love tells us absolutely nothing about whether a particular relationship is healthy or good for us. Steamy starts are compelling, but intense emotions can block our objectivity and blur our capacity for clear thinking and clear speaking.

A friend, Amy, fell in love with a woman at a weekend workshop on gay activism, and she's already reorganizing her whole life to be with her. Amy has known this person for only three weeks, yet she's planning to give away her beloved cat because of her new girlfriend's allergies. Amy's friends are skeptical and want her to slow down, but she's convinced she's found true love.

And maybe she has. Love means something different to every person who feels it, and it has no God-given definition. After all, if Amy *feels* in love, well, then she's in love—no matter what anyone tells her. Some people do experience a profound and immediate connection to another person that proves to be enduring. But intense feelings, no matter how consuming, are hardly a measure of true and enduring closeness. Intensity and intimacy are not the same thing, although many people confuse the two.

It doesn't matter whether Amy calls it *love* or *sauerkraut*. The most important question is not the intensity of the love we feel, but whether the relationship is good for us and whether we are navigating our part of it in a solid way. Time and conversation help to size this up. Is there a sense of safety, ease, and comfort in the relationship that makes authenticity and self-disclosure possible? Does the person we love enlarge (rather than diminish) our sense of our self and our capacity to speak our own truths? Is the connection based on mutuality, including mutual respect, mutual empathy, mutual nurturance and caretaking? Are we able to voice our differences to bring conflict out in the open and resolve it?

Only when we stay in a relationship over time and evaluate it with both our head and our heart can we begin to put it to the test. This involves lots of talking so that hot spots—whatever they may be—can be spoken to openly. Discussing differences up front is no guarantee against future problems, but it can help both partners assess their ability to negotiate, consider each other's feelings, and compromise when appropriate. Dealing with differences will put the clarity of your voice—and your capacity to listen to the different voice of the other—to the test.

THE DILEMMA OF DIFFERENCES

While listening to national public radio on the way to work one day, I happened to catch the tail end of a truly chilling interview with a white supremacist. I told myself that he too was one of God's children, and that he surely possessed some good qualities along with his very bad ones. I reminded myself that he was once a cute little baby boy, and that he didn't nurse at his mother's breast or roll down the street in his little stroller plotting how to keep America pure and white. I pictured him giving a great big smile to the nice African-American or Jewish woman who stopped to coo at him, just as he smiled at the nice, white Christian lady. But something happened to this poor little guy along the way to growing up.

Something happens to almost all of us, although, thank goodness, not to the extreme of preaching hate. Humans don't tend to do well with differences. We learn to hate a difference, glorify a difference, exaggerate a difference, deny, minimize, or eradicate a difference. We may engage in nonproductive efforts to change, fix, or shape up the person who isn't doing or seeing things our way. In the history of nations, families, or couples, folks find it hard to discuss their differences in a mature and thoughtful way.

Reassuring Sameness

Of course, there's something to be said for huddling together with folks who are just like us. It's easy to voice your thoughts and feelings to someone who already agrees with your every word and is going to nod vigorously with approval as you speak. Tempers are less likely to flare when two people think, react, believe, and vote exactly the same way.

In contrast, differences—even minor ones—can drive a wedge between people. But like it or not, differences will inevitably emerge in any close relationship, and thank goodness for that. What could be more boring than hanging around with folks exactly like ourselves? Differences don't just threaten and divide us. They also inform, enrich, and enliven us. Indeed, differences are the only way we learn. If our intimate relationships were composed only of people identical to ourselves, our personal growth would come to an abrupt halt.

GETTING PAST THE VELCRO STAGE OF INTIMACY

In the early stages of a relationship, partners tend to overlook, excuse, or submerge their differences, or they're taken by the novelty and find the differences exciting or appealing. If you just happen to be in the Velcro stage of a romantic relationship, you're probably in a trance. As a French proverb tells us, "All beginnings are lovely."

When you're overly eager for a relationship to work, you will resist getting differences out in the open, looking them straight in the eye, and having a good fight when necessary. Instead, you may ride the relationship like a two-person bicycle that will topple over if there's not perfect agreement and togetherness. The urge to merge is very strong, and if you're in the grip of it, you'll submerge the clarity of your voice in order to preserve a positive picture of the other person and the relationship.

Intimate beginnings pull for a pseudo-harmonious togetherness. That said, I encourage you to resist that pull and to be as clear-eyed and awake as possible. Do as much talking and listening as you can before you entwine your emotional and financial futures. You can't choose your kids, parents, or relatives, but you can choose your partner, preferably after a courtship (to use an old-fashioned word) that allows you to enlarge and deepen the conversation.

Intimacy usually develops among people who share deeply held beliefs and core values. But closeness should not be confused with sameness, and relationships go best when we get past glorying or devaluing differences and face them with curiosity and respect. Dealing with differences is the perfect training ground for making choices about how we speak and how we listen. A relationship built on silence, on the shedding, or suppression of differences, doesn't have a strong foundation. Nor does it help for two people to get polarized around differences and divide into opposing camps.

WHEN FRIENDS BECOME ROOMMATES

Close friends tend to voice differences and negotiate solutions more easily than partners or mates. After all, friends can retreat to separate spheres and adopt a "live and let live" attitude. Also, friendship rarely becomes a nest of extreme pathology; if we consistently feel diminished, silenced, or unheard, we don't just dream of escape—we get out.

Once under the same roof with a partner—hearts, finances, and futures intertwined—it's harder to clarify where we can compromise, give in, and go along—and where we can't. So let's start with friendship for a model of how to speak to differences and negotiate them.

An incident with my longtime friend Judy Margulis comes to mind. As college freshmen, Judy and I lived in the same dorm (actually, a big old house) when we both arrived in Madison, Wisconsin. Judy was neat and orderly, while I was messy. This difference wasn't a problem until I asked Judy to room with me the following year.

You won't be surprised to learn that Judy's neatness didn't pose a problem for me. I would never have taken offense even if she wanted to spend all her spare time tidying up after me. But my habits did pose a problem for her, which, you may have noticed, is how this particular difference generally works out. I've yet to hear the messy person complain, "I'm so upset! I throw my clothes on the floor, and when I come home that inconsiderate partner of mine has picked them up, folded them, and put them neatly away!" Nope. Instead, it's the neat person who tends to get grumpy and irritable.

Actually, that's not true 100 percent of the time. My friend Marcia was not pleased when her college roommate rearranged her books by size and color. Similarly, many of us would feel outraged if a partner tidied up our out-of-control study or work space without our permission. There's the matter of privacy, personal space, and our right to have others stay out of what we take to be "ours." That said, I've been a therapist long enough to know that it's usually the person who messes up the shared space who is viewed as the problem.

Judy handled her dilemma about rooming with me in an exemplary way. She told me that she'd room with me only if I would agree to keep the room neat on a daily basis. She didn't ask me to sign a written contract, but she was *that* clear. She wasn't just sharing her thoughts and feelings. She really meant it. I negotiated one loophole, which was that I could throw stuff on my bed if I kept everything else neat. Judy agreed, and we roomed together happily the following year.

Judy just naturally followed all the rules of good communication. She didn't criticize me or put me down. She didn't imply that her neatness was a virtue (although she may well have believed that) or that my disorganization was a sign of personal failure. She didn't talk about our differences in terms of good or bad, right or wrong, or better or worse. She appeared uninvested in changing or fixing me, nor did she take it personally that my belongings usually landed first on the floor. She simply clarified that she wouldn't live in a messy room, and she let me know specifically what she needed from me in order for us to be roommates.

Part of the reason Judy was able to express herself so clearly, is that it was a low-pressure situation. If we hadn't been able to come to an agreement, we'd have remained friends but we would not have shared a room. Let's look at this same challenge for an intimate couple where the stakes are higher and the emotional field more intense.

I'M NEAT, HE'S A SLOB

Mona sought my advice because she was "a total neatness freak" and her boyfriend Dan was an incredible slob. She was the sort of person who lined up her shoes in a perfect row and never went to bed with even one dirty spoon in the sink. In contrast, Dan messed up every room he walked into and failed to notice that anything needed to be picked up. Mona put the problem this way: "We're discussing marriage, and he says he can't change. It doesn't make sense to leave him over this one issue, but I'll go nuts if we're living in the same apartment and I'm always picking up after him." She wanted my advice and was curious about what I'd do and say in her situation.

Of course, both Mona and Dan might benefit by becoming more like each other. It would be great if she could loosen up and become less of a "neatness freak," and if Dan could do a better job of picking up after himself. After all, two mature, kind-hearted friends, forced to share a common space, would likely accommodate each other. Why should two people who presumably love each other do any less? Unfortunately, the challenge of intimacy tends *not* to evoke our most mature selves. Instead, *she* anxiously pursues him to change, and *he* stubbornly digs in his heels in response to her efforts to change him. So nothing changes at all.

What to Do?

What can Mona do? First, she needs to be clear about who owns the problem. Dan is as happy as a lark with his customary ways and can easily ignore Mona's neatness. Mona, for her part, is upset by his behavior. Quite simply, she has the problem, which isn't to imply that she's wrong, to blame, or at fault in any way. Mona needs to resolve her current dilemma herself, because no one else will do this for her.

Mona needs to keep talking to Dan about her problem ("I'm a neatness freak") without acting like it's her job to change him. It will help matters if she can avoid the usual communication stoppers, such as criticizing, lecturing, admonishing, threatening, analyzing, and blaming. She can stick to "I-language" (non-blaming statements about the self) as she tells Dan what she's feeling and what she wants from him. For example, "Dan, I need to tell you something about myself. I'm a neatness freak. I get really anxious when things around me get out of order. I can't handle it. My nervous system starts twitching. If we're going to live together, we have to make a plan, or I'll go nuts."

Making a relationship work obviously requires good humor, generosity, a tolerance for differences, and a willingness for give-and-take. Slobs deserve a room of their own (or a corner or a big armchair, depending on what space permits) where they can dump stuff in a big heap. Every neatness freak deserves to have some rules about public space. There is always room to negotiate in a relationship—and if there isn't, we need to reconsider the relationship.

But what's at stake here is much more than the level of tidiness in Mona and Dan's apartment, should they share one. The real issue is how they can talk about their differences, how well they can hear each other, and how committed they are to finding a solution they both can live with. The worst-case scenario is that they'll each dig in their heels and get stuck in an endless cycle of fighting, complaining, and blaming. The best-case scenario is that they'll both be flexible, creative, fair, and sincere in their attempts to consider each other's feelings and hear each other's voices. If they limber up their brains, they'll find a way to resolve the problem—unless it runs much deeper than the issue at hand.

What would I do in Mona's situation? I wouldn't break off a relationship because my boyfriend was a slob. But I would end it if I felt that the other person didn't consider my feelings, refused to change behavior that was obviously painful to me, and pleaded helplessness ("I can't change") in response to reasonable and fair expectations. I'd also take a good hard look at how my own communication (like nagging or being critical) might be contributing to the problem. I wouldn't pick up after him unless I could do it quickly and without resentment, say, sweeping through the house and dumping all his stuff on his one big armchair. Maybe I wouldn't clean as often or as thoroughly as I used to. I'd also be happy to make a deal. For example, I'd do double duty on cleanup, and he'd do a double shift of shopping and cooking. Couples do best when they can lighten up and accommodate each other.

Mona is smart to be struggling with this particular issue sooner rather than later. Marriage (a subject we'll get to later) tends to make it more difficult for two people to really hear each other and negotiate their differences. Gloria Steinem once urged an audience to do one outrageous act a day, and she suggested as a start that women say to the man they live with, "Pick it up yourself." Someone in the audience agreed that was a fine idea, but how do you make him do it? A small, elderly lady in the back of the large hall caught Steinem's attention and spoke out. "I nail his underwear to the floor," she said. "Nail as long as you need to!" advised a second woman, "and then get the floor done over." Of course, our voice is arguably a preferable tool to a hammer for driving a point home, but some people take drastic measures when they don't feel heard.

For some of us this particular issue ("I'm neat, he's a slob") would be a minor one, along the lines of "I like vanilla ice cream, and he likes strawberry sorbet." But for Mona, it was right up there with the big ones—He wants children and I don't, or I love sex but he's not interested. For this reason, Mona should take all the time she needs to see if she and Dan can find a solution they both can live with comfortably. She shouldn't give up her own living space until she's clear about where Dan stands and what's acceptable to her.

Maybe Dan will decide to neaten up, at least to some extent, although clearly he'll never meet Mona's standards. Or, alternatively, Mona may decide that she's going to find a way to live with Dan's messiness, because Dan is flexible and fair on other important matters. They may try living together to see how it goes, or they may decide to marry and keep separate places. But if Dan is clear—whether through his words or his behavior—that this aspect of his behavior won't change, Mona should clarify to herself, and then with Dan, what she can and can't accommodate. She should not decide to marry him and then complain about it.

We need to know what we're looking for in a partner, and we should never believe that our love (or nagging) has the power to create something that wasn't there to begin with. Nor should we ever marry for love alone. We need to examine our core values and beliefs (what *really* matters to us in a partner), so that we know where we can compromise and where we can't.

How we each compromise is a deeply personal matter. No one else can know whether you should disqualify a potential partner because he or she lacks something important to you: money, good looks, reliability, humor, personal hygiene, an erotic imagination, warmth, a love for the outdoors, or tidy personal habits. Such decisions ("Is she the one?" "Am I compromising too much?") can be excruciatingly difficult to make, but no one else can do this work for us. The clarity of our voice rests on the clarity of our self-awareness regarding what we want and feel entitled to, and what we are willing to settle for. It can take time, patience, conversation, and silent meditation or reflection to sort this out.

WHAT ARE YOU LOOKING FOR?

Most of us think we're clear about what we want in a mate. While individual tastes vary, we want a partner who is mature and intelligent, loyal and trustworthy, loving and attentive, sensitive and open, kind and nurturing, competent and responsible. I've yet to meet a woman who says, "Well, to be honest, I'm hoping to find an irresponsible, distant, ill-tempered sort of guy who sulks a lot and won't pick up after himself." Or, "Hey, can you fix me up with that cute friend of yours—you know, the one who is totally self-absorbed and conversationally impaired."

And yet few of us really evaluate a prospective partner with the same objectivity and clarity that we might use to select a major purchase. We wouldn't buy a used car off the lot just because it looked great and felt really comfortable to drive. We'd check out its history and ask for the facts, with our radar out to detect dishonesty or hype. We might consult with a clear-thinking, car-savvy neighbor. And we'd enter the negotiation with a few criteria of our own that were deal-breakers—maybe air conditioning, good mileage, or safety features such as antilock brakes.

We need to be at least this careful in matters of the heart. In the name of love, we may lower our standards, silence our questions and concerns, and even abandon our friends for someone we probably shouldn't trust to water our plants when we leave town. We may keep sleeping with someone whose behavior is equivalent to waving a big red flag in our face.

Of course, in picking an intimate partner, we don't compile a list of key criteria and then proceed with the selection process in a totally intellectual fashion. But it's not a bad idea to consider this approach. Some women have found it helpful to jot down the top five traits, qualities, and behaviors that are important to them in a life partner ("financially stable, reliable and responsible, talks about problems, shares household responsibilities, likes my kid") and then to refer to their list in evaluating the relationship in question.

We need to keep reaching for the facts through conversation and observation. It's data if your boyfriend won't visit your family or if he tells you that all his previous girlfriends and ex-wives were big losers. It's data if you stop voicing your wants, expectations, and questions because you're afraid to put him to the test. No one fact or combination of facts should make us reject someone if we have a really good feeling about that person and what's happening between us. But we need to keep moving in the direction of becoming more real ourselves and more objective about our partner.

THE VOICE OF THE BODY

Words aren't the only ways we get information about each other. Talking is essential, along with observing whether the other person's talk is backed up by responsible action. But we also learn a lot about another person through the full range of our senses. Choosing an intimate partner is not a purely intellectual matter but also a task of the heart, which goes beyond language to involve feelings, desire, chemistry, and intuition.

In truth, we come to truly know the other not only through words but through an intuitive understanding or "reading" that comes through the body. We know, through our bodies, whether a particular interaction leaves us feeling energized, uplifted, and inspired—or the opposite. We know, through our bodies, whom to trust, or believe, or avoid. What we call intuition and "gut reactions" is shorthand for the extraordinary human capacity to process information about another person that is beyond words.

Recently I sat for twenty minutes at a bus terminal in Providence, Rhode Island, and watched a small group of young men and women communicating in American Sign Language. I found one of these guys so appealing that I wanted to kidnap him and take him home to Kansas to be my friend. Okay, maybe he was a bigot who was telling his friends how upset he was at the prospect of the government banning his assault rifles, but I doubt it. Nor was I responding to anything as obvious as body language, good looks, poise, or grace. We are constantly taking in nonverbal information about people that we sense automatically and effortlessly.

If a person's words tell me one thing ("I'm feeling close to you") but my automatic knowing intuits something different (I sense distance, disconnection, a "not-thereness"), I put more trust in what I feel than in the words I hear. I know when another person is distracted, even when that person claims to be paying attention. I make automatic judgments about who is kind, trustworthy, and forthcoming and who is not. Of course, we all make mistakes (we think the other person is standoffish when actually she's shy), but it's interesting to think about how automatically we can get a sense of the other person that doesn't rely on the *content* of what is said. To read other people with some accuracy we need to feel comfortable, safe, and relaxed in their presence—and to trust our gut when we don't.

IS HE LYING TO YOU?

When I was advice columnist for *New Woman* magazine, a thirty-three-year-old executive wrote me that she couldn't get over her suspicion that her fiancé was having an affair. For many months she had sensed that he was distant and "not there." She had a strong voice, and she confronted him repeatedly, only to be told it was all in her head. Two of her married friends had secretly hired private detectives in similar circumstances and were encouraging her to do the same. She wanted to know if I agreed with her friends' advice.

In my response, I told her that, personally, I wouldn't do it. If I hired a private investigator, I'd feel as if *I* were the one having the affair. I don't think I'd be able to restore intimacy, to say nothing of my self-regard, if I introduced this level of secrecy or deception into a relationship. I wouldn't want to respond to mistrust by becoming untrustworthy and secretive myself. Nor would I stay with a partner who had invaded my privacy in a serious way, say by tapping my phone, bugging my office, or hiring someone to trail me. If I was really convinced that I needed a detective to learn more about a person, I wouldn't marry him.

I encouraged this woman to trust her intuition that something was wrong and to take whatever time she needed to deepen her knowledge of this man and to restore her level of comfort. How well did she know his family and friends? What questions had she asked about his past relationships? What was his track record on openness, honesty, and fidelity? Was he usually open or closed about difficult emotional issues in his life? Did she consider herself a jealous person, with a history of overreacting and reading too much into things?

I also suggested that she keep giving voice to her concerns with her fiancé. She needed to say, "I feel that something has changed in our relationship, and you keep telling me it's all in my head. But my feeling that something is wrong isn't going away." If she isn't satisfied with his responses, she needed to take the conversation to a deeper level: "Look, I keep sensing a funny distance, and I worry without knowing what to think. I wonder if there's another woman. But maybe you're in trouble at work, or you have a health problem, or you're questioning our engagement. Whatever it is, I want us to be able to talk about it. The hardest thing for me is to sense that something is wrong and to be told I'm just imagining it."

She can voice her concern strongly, but not accusingly—a challenge in all intimate relationships. She can also put her marriage plans on hold and take whatever time she needs to gain more clarity about the level of intimacy, honesty, and trust between them. If she doesn't trust him to tell her the truth, or she continues to sense a funny distance or "not-thereness" from him, it's unwise to marry.

I thought my answer dished up plain common sense, but it proved to be highly controversial. I was surprised by the number of letters I received. Many readers had themselves hired private detectives to investigate their suitors and reported that this move had saved them from a tragic mistake. "You've been living in Kansas too long," one woman wrote me. "You're as naive as Dorothy, and your bad advice could cost this woman everything."

Of course, advice is always problematic if it's given or received as anything more than just one perspective that may or may not fit the other person. An advice columnist skates on thin ice, having a short space to respond to a complex question offered in good faith by a total stranger. I trust my readers to be the best experts on their own selves—to run with advice that fits and to ignore the rest. But I'd still say the same thing today in response to this woman's dilemma.

Nor am I as naive as my coastal readers may think. I'm well aware that no matter how awake we try to be, we can't prevent another person from lying to us, whether through words or silence. Nor can any of us know for sure whether a partner is telling the entire truth. The best we can do is to stay alert, live our lives honorably, and speak truly. In my mind, hiring a private detective is a step in the wrong direction.

STAND BY YOURSELF

The most important voice we need to trust in a relationship is our own. We need to trust ourselves to perceive and process important information. We need to use a clear, strong voice to bring our knowledge of the relationship into sharper focus and test out what's possible, rather than comfort ourselves with fantasies about how our partner might change in the future. We need to speak up and insist on fair treatment and respect.

If we can't do all these things, it's difficult to fully trust the other person or the future of the relationship. If we *do* trust ourselves, we'll know in our gut that there are some behaviors we just won't tolerate and some places we just won't go—even if we do make an initial poor choice.

Setting Limits up Front

Consider Lavetta, who is dating the most appealing guy she's met in years. She spent the first few months of the relationship on that proverbial cloud nine, but as they have become more involved, he has been acting more possessive. They have a serious difference of opinion about her putting time into other relationships. He wants her to quit her reading group because it meets in the evening, he sulks when she goes to the movies with her friends, and he's pressuring her to cancel camping plans with her two sisters because he'll be off work at the same time and wants her to vacation with him.

When Lavetta tells him how she feels, he seems respectful, but she seeks my help because she's worried about his increasing displays of possessiveness. "I really don't want to dump this guy," she tells me, "but I'm freaked out by his behavior."

Lavetta doesn't need to dump him, and she doesn't need to freak out about their different perspectives. But she does need to keep a close eye on *her* behavior to make sure she doesn't start giving up her life for this guy. His possessiveness will only get worse if she goes along. So Lavetta should not cancel her reading group or movie dates or camping plans if doing so doesn't feel right to her. She can say no—and aim to find other ways to make him feel important and special. She can keep the lines of communication open to leave room to hear his feelings and concerns.

Lavetta needs to hold fast to her independence, a trait that is important to her. This doesn't mean being rigid and uncompromising or failing to consider her boyfriend's wants. All relationships require compromise and give-and-take. But having a voice also requires us to decide what values, beliefs, and priorities are non-negotiable. If Lavetta takes care not to lose herself in this relationship, one of three things will happen. Either her boyfriend will rise to the occasion, maturing a bit himself, or he'll dump her and hook up with a woman who will join him in the urge to merge. Or Lavetta will get the clarity she needs to end this relationship.

The challenge in all intimate relationships is to preserve both the "I" and the "we" without losing either when the going gets tough. If we're faced with a choice, we need to choose speech over silence, keep our behavior in line with our stated values and beliefs—and save ourselves first.

I should add that we learn a great deal about a partner by observing him among both our friends and family and his. In contrast, you won't know him or her well, no matter how much you talk, if you insulate your relationship from other key relationships. So you need to do a lot of talking (and listening, observing, and thinking) in varied settings and situations to refine the process of knowing the other person and being known.

THE HONEYMOON IS OVER

Marriage—or living under the same roof—often solves the problem of the Velcro stage—denying, excusing, or sentimentalizing differences. The next stage of the relationship finds two people angrily polarized around the differences they failed to notice or initially found appealing. They may dig in their heels with little flexibility to openly consider the other person's point of view and honor a reasonable request for a behavioral change, confusing angry exchanges with "being real" and having an authentic voice.

Whether you've been part of a couple for four—or forty—years, you will still have to face the challenge of differences. Long-term relationships suffer when we don't face differences with tolerance, humor, and respect. They also suffer if we become so tolerant of differences that we expect too little from the other person, or settle for unfair and compromising arrangements that erode our sense of self.

There are two essential *voice challenges* in a committed, intimate relationship such as a marriage. First, we need to clarify a bottom-line position and stand behind it. Second, we need to speak to the positive in the other person and the relationship and to warm things up. Let's look carefully at these approaches, both of which are essential for a relationship to work out well.

CHAPTER 8

Marriage: Where's Your Bottom Line?

My husband, Steve, was after me for a while to keep the bedroom neat, not to throw my clothes on the floor when I climbed into bed at night, or at least to hang them up in the morning before I went to work. He'd complain periodically, and I'd shape up for a while before lapsing back into my old habits. I knew that Steve was "just complaining." He wanted me to change, but it wasn't *that* big a deal to him. When Steve *really* means it—whatever the "it" may be—I know it.

I know, for example, that I can't treat Steve in a rude or disrespectful way, without apologizing and getting a grip on my bad behavior. Of course, I'm ill-tempered and obnoxious every now and then. But if I made a habit out of it, I wouldn't have a marriage. It's not that Steve would announce, after three decades, "Harriet, if you keep treating me in a rude fashion, I'm filing for divorce." He wouldn't have to. I know that Steve is in our relationship for the duration, that leaving is the last thing on his mind. I also know that if I became a rude, disrespectful, or unfair partner—if, over time, I stopped allowing Steve's pain or voice to *affect* me—my marriage would be over. I know this because I know Steve.

Here's an example. There was a time when I was working too hard, traveling too much, and feeling overly responsible for bread-winning. I got grouchy about it and took jabs at Steve. To his credit, he let me know that he wasn't going to listen to my criticism about his not working enough hours to suit me. When I "started in on him," as he put it, he'd lift his hand and firmly say, "Stop. I don't want you to talk to me in that tone of voice." If I didn't stop or change my tone, Steve would end the conversation right then and there. He'd say, "I don't want to be criticized, and I won't be talked to that way." Depending on his mood, he'd make his point with great maturity or great immaturity. Either way, he had a clear bottom line about what sort of conversations he wouldn't participate in.

Steve was not stonewalling me or ruling this (or any other) subject off limits. He expressed a willingness to sit down with me and look at both of our work schedules. He was clear that he was available for a conversation on any subject. He insisted only that I approach him as if he were a collaborative partner and not a big screwup. That was his bottom line.

Steve also knows when I really mean it, even about little things. After he reorganized his study one Saturday, he left a couple of big boxes in our kitchen rather than putting them in the basement. The boxes weren't blocking traffic, but I didn't want to look at them, so on Sunday I asked him to put them away. When Monday night rolled around, Steve still hadn't done it. So I said, "Steve, please take your stuff to the basement. It's really bothering me."

To my credit, I made a *specific behavioral request* ("Please take your stuff to the basement"), which is a hallmark of good communica-tion. Often, when we're irritated, we go beyond asking a partner to change specific behavior, and instead broaden our statement to include a jab at the other person's character or to bring up past sins.

For example, it would be inflammatory to say, "Steve, you're so inconsiderate. You don't think about other people. You never follow through. Please take your stuff to the basement. It's really bothering me."

We've probably all made our fair share of such global comments, which are a recipe for fighting rather than for being heard. But communication skills are not my focus here. Instead, I'm talking here about something more central than the words we choose. The point about Steve's moving the boxes is that *I really meant it*. A few boxes in the kitchen are inconsequential, but being ignored in response to a simple and perfectly reasonable request is a big deal to me. So if the boxes were still there on Tuesday, I'd find a way to take the conversation to a new level. For whatever reason, the boxes in the kitchen were much more of an issue to me than a neat bedroom floor was to Steve.

If I needed to take the conversation further, I'd try not to speak out of anger or to overreact. For example, I wouldn't put my hands on my hips and announce sternly, "Steve, I'm on strike until you move those boxes. Don't count on me for *anything*." But at the very least, I'd need to know when he planned to move the boxes in order for me to sit down and enjoy dinner with him, go to the movies, or continue business as usual. And I'd need him to know how much it was upsetting me—not so much that the boxes were still there, but more importantly that he was choosing to ignore my feelings.

I don't mean to imply that Steve and I resolve every difference by taking a bottom-line position with each other about what we can or can't live with. Over the decades, we've been locked into some rather dramatic, nonproductive fights where each of us keeps doing more of the same. Sometimes my attempts to change Steve around a particular issue have only made things worse. Lightness and humor usually get us through both the small differences and the difficult impasses. When they don't, we find some other way of negotiating or tolerating our differences.

That said, we both know there is a line we can't cross, that there are certain behaviors the other won't tolerate over time. Even when it's not spelled out in words, couples know each other's bottom line, just like kids know what they can and can't get away with. And like a kid, a partner may keep pushing the limits until the other person says, "Enough!" and really means it. That place is our bottom line.

ESTABLISHING YOUR BOTTOM LINE

Clarifying a bottom line is perhaps the most difficult challenge of finding voice and being heard. A true bottom-line position is not an ultimatum. It is not a threat or a reactive position impulsively decided on at a time of high intensity ("Damn it! If you do that one more time, I'm leaving!"). It is not an expression of desperation or a last-ditch attempt to force a partner to shape up. It is not a mixed message, where our words say one thing ("I can't continue to take this") and our behavior says another (we continue to take it). Instead, a bottom-line position evolves from a focus on the self, from a deeply felt awareness—which one cannot fake, pretend, or borrow—of what we need and feel entitled to, and the limits of our tolerance. We clarify a bottom line, not primarily to change or control the other person (although the wish to do so may certainly be there) but rather to preserve the dignity, integrity, and well-being of the self. Everyone is different, and there is no right bottom line for every person. But if we have no bottom line on important issues (we complain, but nothing changes and we don't follow through), relationships—and our sense of self—spiral downward.

A bottom-line position stems from deeply held values and gut-level responses that define what we can accept and still feel okay about in a relationship, in the other person, and in our own self. When we take a bottom-line position in any relationship ("Dad, whenever I see you've been drinking, I'm going to get up and leave"), we need to clarify that we are acting for our self rather than against the other person ("Dad, I'm doing this because I love you and it's too painful for me to be around you when you've been drinking, and especially because it reminds me that I may not have you around for a very long time"). The first step to establishing a bottom line is self-knowledge and from there, self-expression.

Stand Like a Mountain, Bend Like Grass

A bottom line need not be rigidly held, but rather can be reevaluated in light of new experiences and information. For example, a client of mine was living with a man, Simon, who kept balking at setting a marriage date. When she first came into therapy, marriage was her biggest priority, even at the expense of losing her partner. So she took a bottom-line position and said, "Simon, I love you and I want to spend my life with you. But marriage is so important to me that I need a commitment. I know you need more time to make a decision, but let's decide how much time." To her credit, she clarified a bottom line not as a threat or attempt to rope him in, but rather as a calm clarification of what she could accept and tolerate. She was clear about her limits, saying she could wait another six months, but if he was unsure after that time, she'd move on, painful as that would be.

When the agreed-upon time period had passed, Simon was still not ready to commit to marriage. At this point my client decided not to push him. She didn't want him to be at the altar under pressure, feeling it was not his heartfelt choice. But as she contemplated packing her bags, she realized that she wanted to preserve this relationship and that she could enjoy their time together even without the commitment she wanted. She worked successfully at letting go of her focus on marriage as an end in and of itself. She also learned to underreact—rather than overreact—to Simon's expressions of ambivalence about marriage. Yet this change in her attitude was not "caving in" or a violation of her values and priorities. Rather, she reevaluated them in light of the facts, and made a new decision that she never regretted.

In sum, a bottom-line position is firmly but not rigidly held. It's nonnegotiable unless new learning challenges us to reevaluate our core values and beliefs. Such a reevaluation is not the same as going along with someone else's program at the expense of the self.

If we're confident about our ability to hold firm when we need to, we can be loving, curious, and connected as we voice our needs, expectations, and limits. We can listen to another person with an open heart and try to understand a different point of view, rather than delivering ultimatums.

Live and Let Live?

We can strive to adopt a live-and-let-live attitude with friends or with family members who live separately from us, but it's much more difficult to respect differences with an intimate partner living under the same roof. When two lives are financially, emotionally, and logistically intertwined, differences can severely affect one's current and future well-being.

Some of us think about our bottom line only when we're at the crossroads of staying or leaving. In reality, it's there with us every day in every important relationship. It doesn't matter whether we believe that the other person's behavior has been caused by illness, environment, attention-deficit disorder, bad genes, slothfulness, evil spirits, or an overzealous wish to be helpful. If we have no bottom line in a relationship, then that relationship is bound to become increasingly impaired.

If we're looking at someone else's marriage, the "right" position to take on important differences may seem obvious and easy. But from the inside, establishing a bottom line may feel like utter confusion. We may start out clearheaded in the conversation, only to find our brain turning to mush when the other person ups the ante and puts us to the test—which will inevitably occur. And if it's hard to say, "This is what I need to do for myself," it's all the more difficult to say, "This is what I expect from you," and then, if we're ignored, to decide what to say next and how to say it.

Moving forward in a couple relationship requires one thing above all else. We need to stop focusing on how impossible our partner is, and instead focus relentlessly on the clarity of our own voice—the conversations we have or avoid, the positions we take or fail to take, the places where we stand firm or cave in. Let's look at how difficult this challenge can be in a troubled marriage.

AN "IMPOSSIBLE" HUSBAND

My friend Grace accommodates enormously in her marriage. On important matters she gives in, goes along, buckles under. She submits to circumstances that she believes are unfair, and accepts behaviors she wouldn't tolerate in a roommate or best friend. Her accommodating behavior is deeply problematic, because it violates her core values and beliefs about equality and mutuality in relationships—not to mention that it ensures that her marriage will continue as is, highly problematic and unsatisfying.

Somehow Grace and I got to chatting about those big boxes Steve left in the kitchen when he was reorganizing his study. Grace tells me that her husband, Marshall, would leave the boxes there "forever." She describes Marshall as "passive-aggressive," "very controlling," and "a piece of work."

"You're lucky that Steve is such a nice guy," Grace comments. "If Marshall left boxes in the kitchen, nothing I could say would move him. The more I made a big deal of it, the more he'd dig in his heels. Sometimes I ask Marshall to do something, and he sits there and says *nothing*. He doesn't even answer me!"

"I couldn't tolerate that," I say. "I'd be walking around feeling angry and frustrated all the time."

"I do get angry and frustrated." Grace responded, "but aren't you the one who teaches that you can't change another person who doesn't want to change?"

"I'm not talking about changing Marshall," I explain. "I'm saying something about myself."

"Are you also saying something about *me?*" Grace asks.

"Well, yes," I say. "You're a much more tolerant person than I am. Like I said, if Steve behaved like that, I couldn't tolerate it."

"Marshall isn't going to change," Grace reiterates.

"I'm not talking about Marshall," I say lightly. "I'm commenting on a difference between us. You have the tolerance of a saint."

Grace laughs. "No, I don't think I'm a saint."

"Saint Grace," I tease her lovingly.

Marshall, indeed, behaves badly in their marriage. Grace invites me to see him as a terribly difficult person, although she's also deeply loyal to him. I do see him as difficult, but in my conversations with Grace, this isn't my focus. Instead, I focus on Grace. She's the one in the couple I'm close to, and she's the one who is taking it. If I join Grace in her single-minded focus on Marshall as the problem—rather than pointing out how accommodating Grace is— I couldn't be helpful to her. It's rarely constructive to join a critical bandwagon about another person's husband.

"Maybe I have no bottom line?" Grace suddenly announces, more as a question than as a statement of something she really believes.

"Everyone has some bottom line," I say. "What if Marshall started sleeping with his ex-wife? What if you told him how painful it was for you but he still wouldn't stop?"

Grace was silent.

"What if he hit you?"

Here Grace didn't miss a beat. "I come from an abusive family. I wouldn't tolerate being hit."

"I know you wouldn't," I said, "and you really mean it. It's not just words, it's a position in your gut. It's a place where you won't be violated, where you won't compromise. Marshall knows that he can't hit you and expect to keep his marriage. That's a bottom-line position for you. We all reach a point where we say, 'Enough!' and really mean it."

Grace was thoughtful. "Are you saying that for you, Steve's not taking the boxes to the basement is the equivalent of Marshall having an affair or hitting me?"

"No, of course not," I say. "It's not that the *behaviors* are equivalent. But if Steve was doing something unfair or hurtful, and if he refused to respond to my complaints and distress, I'd be so angry that I'd be a total mess. So I'd stay in the ring and keep moving the conversation to another round until he heard me and something changed. I'd have to, because I just couldn't take it. But that's me. You and I are different."

"Are you feeling critical of me?" Grace asked, after a long pause.

"No, not critical," I say. "But I worry about you. I worry that you're going along with behavior that's occurring at your own expense and that's not good for you. I also don't think it's good for Marshall or for your marriage."

I really do believe this—that we diminish the other person when we tolerate behavior that diminishes the self, when we don't expect enough from the other person or put him to the test of what's possible. Marshall can only lose self-respect when Grace tolerates disrespectful behavior from him.

My Dilemma with Grace

I also face a challenge myself with Grace about speaking up or staying silent. We've been close since our freshman year of college. We speak easily and intimately no matter how great the geographical distance between us. But how do I find *my* voice with a dear friend who has no effective voice in her marriage?

It's very hard to watch someone we love disappear into a marriage. Only very highly evolved persons can watch their best friends or family members behave like total neurotics or lose themselves in a relationship and not feel compelled to yell at them or whip them into shape. But having a voice doesn't mean that we're free to offer unsolicited advice to folks because we feel driven to confront them with our ultimate wisdom. Since I happen to have this particular problem, I need to keep reminding myself that it's not my job to tell Grace how to live her life and conduct her marriage.

I also recognize that Grace may have more at stake in maintaining the status quo—or more at risk in challenging it—than I can begin to appreciate. We can't really know what's best for another person, or how much change anyone can tolerate at a particular point in time. And we never really know what anyone else's relationship is actually like from the inside.

Staying silent isn't the answer, either. I love Grace, and I want to talk with her about everything, troubles and joys, so I continue to find some way to speak up. Sometimes I just listen when Grace tells me about Marshall's irresponsible actions. I might say, "Oh, Gracie, honey, that sounds really hard." Or I'll tease her, "Grace, how come you haven't gotten that man to shape up in ten years! What's wrong with you, gal?" Sometimes I ask serious questions. "Grace, if nothing changes between you and Marshall, where do you see yourself five or ten years from now?" Or, "If you did leave Marshall, how do you imagine you'd do on your own?" Sometimes I speak to the differences ("Grace, I could never have gone along with that . . .") as a way of communicating that we all have choices.

If all I could do is hold my tongue with Grace about her behavior—or the pattern I see in her marriage (he dishes it out, she takes it)—our friendship would become distant. Intimacy and honor are sacrificed in a close relationship when we withhold our genuine response. Grace would also probably sense my feelings whether or not I said anything, or she'd be left guessing about what I thought. She'd also sense my feelings if I spoke to the differences in a tone that implied that Grace was a loser while I was a superior, more evolved person who would never allow myself to be in her pathetic situation. Shaming and blaming people isn't a good way to teach anybody anything, and it's not a path to deeper, more honest conversations.

The less I need a particular response from Grace, and the more respect I have for her struggles, the more creative I can be about sharing my observations and concerns. I speak openly with Grace not because my words change or even affect her when it comes to her marriage. I speak up because I love her and I want to be real with her. I speak up because it's important to me to do so.

We can't fix another person's problems, but this awareness needn't lock us into silence or serve as our excuse to keep quiet, stay away, or ignore the seriousness of what's happening with a person we're close to. I may never understand why Grace tolerates behavior in a husband that is so incongruous with her stated values and beliefs. I can't know how her part in this pattern serves or protects her, and when, if ever, she will be ready to change it. I do know that Grace's struggle is one that many women will relate to—she may not truly value and regard herself, she has difficulty clarifying the limits of what she will tolerate, she doesn't expect much from men, she compromises too much of herself in relationships, she's terrified and feels guilty to even contemplate leaving, and she takes a certain pride in her ability to endure.

An Ongoing Struggle

The "voice challenge" Grace faces with her husband is far more difficult than the challenge of what I say to her as a friend. Whether Grace protects or protests the status quo, she has so much at stake. From the outside, we can look at another person's marriage and say, "Why is she so wimpy? Why can't she hold her ground?" But the interior of that marriage is entirely different. Something happens when people are caught up in the emotional force field of marriage or coupledom. As a brilliant friend of mine said, "Call it 'marital fusion' or whatever, but whenever I tried to hold my position with my ex-husband, my brain fuzzed over." The point is not that we should give up, but rather that we should develop patience with ourselves, make use of our clear-thinking friends, and get help when we need it.

Grace is still struggling to figure out what her bottom line is in her marriage. I continue to challenge her to work on strengthening her own voice in terms of telling Marshall what she needs and expects. I appreciate that it is far more frightening for Grace to leave Marshall than to stay with him. Grace may need to experience more pain before that balance shifts—and it may never shift.

Grace will take whatever time is necessary to figure out whether she can live with the status quo. If she reaches the point that she just can't take more of the same, she will act and speak from a new and different place. Her bottom line will shift. It won't be a matter of changing Marshall, but of saying, "Look, as I see it, your behavior is hurting our marriage. I can't continue this way and still feel good about myself, or you and our life together."

Should such a point occur, Grace won't get sidetracked any longer by Marshall's protests ("You don't accept me the way I am," or, "You're trying to change me"). Such countermoves on his part are not to the point. Of course, Grace can't change Marshall. Ultimately, he alone will decide how he will behave in their marriage, what he feels responsible for and entitled to, what sort of person he wants to be in the world, and what he will and won't change. Grace alone is responsible for making these same decisions for herself, and for clarifying when she will no longer stay in the marriage because it's too painful or compromising.

CHAPTER 9

"I Can't Live with This!"
Voicing the Ultimate in Marriage

Clara came to see me for brief psychotherapy after five years of marriage to Sam, a computer scientist addicted to drugs and alcohol. Clara had left Sam many times, but she kept going back because, as she put it, "I can't be alone, and I love him." Sam kept promising to get help, but dropped out of every treatment program he started. Shortly before calling me, Clara had turned a corner: she knew with total clarity that she couldn't continue in the same old way.

Clara didn't call me because she was confused, but because she wanted to build her own support network to help her follow through with her plan. She knew exactly what she needed to say and do. During our first meeting, she said, "I'm going to start going to Al-Anon meetings to work on my own healing. I need to tell Sam that I can't stay with him unless he takes action to solve his problem and remains in a recovery program to ensure that he stays clean and sober. I have to find my own living situation right away, and need to consult a divorce attorney for advice."

Even though Clara knew clearly what she wanted to do, she was terrified of taking these steps. She had convinced herself over the years that she wouldn't be able to manage on her own. But deep down she knew that if Sam were hit by a truck tomorrow, she'd find some way to keep going. The knowledge that she could support herself was crucial. Only after we know that we can live without a relationship—and feel entitled to make that choice—can we think, speak and act clearly within it. We can't take a bottom-line position in any relationship—or on any subject—if we believe our safety or survival is at stake.

Clara recognized that saving herself, not Sam, was the task at hand—and that this shift in focus would actually give her relationship the best chance to survive. Several weeks after we started meeting, Clara took a firm position with Sam in a loving, noncritical way. She told him how deeply she cared for him and how much she wanted their relationship to endure. Knowing she was now prepared to act, she found the inner clarity that allowed her to voice the ultimate. She said:

> Sam, I know I've left and returned many times. I need to tell you that I've finally gotten clear that I'm leaving and won't return until you're clean and sober. My intention is not to tell you what to do, but rather to let you know what I can live with. When you drink or take drugs, you're not there for me, and I'm left without a real partner. And because I love you, I can't live here while you keep on hurting yourself. If I stay here while you're doing nothing to help yourself, I know I'll end up feeling like I'm part of the problem. You need to decide whether you're going to get treatment for your addictions. And I need to figure out how to stand on my own two feet and get my own life in order.

Clara spoke wisely and well. But again, communication skills are not the heart of the matter. Even if she had broken every rule of good communication in the book, Clara was ultimately going to be heard. Something had shifted inside her. Finally, she really meant what she said.

Sometimes Clarity Just Happens

How did Clara arrive at a place of clarity, after years of fights, threats, and futile attempts to make Sam stay in treatment? Interestingly, she didn't have a clue about what led to her transformation. She couldn't identify an insight or single pivotal event that led her to say, "Enough!" As she tells it, she left a concert one evening with the knowledge that she was moving out and not returning until after Sam was sober and in ongoing treatment. The depth of this knowing allowed her to speak and be heard.

Clara described herself as having arrived at this new place out of the blue. I imagine, however, that it followed a long accumulation of smaller transforming experiences that went unregistered but together led Clara to shift internally into a new place from which there was no turning back. Perhaps the concert, which she described as beautiful and transcendent, put Clara more in touch with her own potentialities, new possibilities, and capacity for joy, thus making her current situation even less tolerable. But from Clara's perspective, the moment of transformation just sneaked up on her when she wasn't looking. She was, however, ready to listen, speak, and act when that moment arrived.

Clara was able to take her position with dignity and firmness, because she knew that it was right and necessary. Usually we don't have the luxury of this kind of clarity. Marriage is often extremely complicated and riddled with uncertainty. We may feel at a loss. Where do we stand? Should we try to live with something or say we can't? If we *say* we can't, are we really willing to leave? Are we crazy? The following story illustrates the painstaking effort—the nearly endless trying and trying again—that goes into achieving clarity and voice when something as important as the survival of a marriage is at stake.

A HUSBAND'S OTHER "FRIENDSHIP"

Lorraine had been married to Ira for almost a decade when she contacted me for help. Lorraine was Ira's first sexual partner. In the past two years, though, he had become extremely close to a married female coworker named Jo. He assured Lorraine that they were just friends, but Lorraine couldn't get over a "sick feeling" at the mention of Jo's name. She also questioned the friendship because Ira had once kissed Jo passionately at a party when he had been drinking.

On numerous occasions Lorraine tried to convey her feelings to Ira, who insisted she was overreacting. Over time, Ira became hesitant to mention Jo's name for fear Lorraine would get upset. But he still called Jo at home, jogged with her in the park before work, and e-mailed back and forth at night.

Every time Lorraine tried to talk to Ira about his relationship with Jo, he'd reassure her that nothing was going on. But Lorraine didn't feel reassured. Talking to Ira wasn't helping, so Lorraine decided to invite Jo and her husband to dinner, in an attempt to get to know her and defuse the situation. It was a creative idea, but when the evening ended, Lorraine felt even worse.

Several months before Lorraine contacted me for help, she had initiated couples counseling with a different therapist, insisting that Ira join her, which he did. She was hoping for a neutral space to talk about this painful subject and to work on "intimacy issues." This therapy went in circles and wasn't helpful, which didn't surprise me. Couples therapy is a total waste of time if one person is having an affair—including an emotional affair—but not owning up to it.

When the therapist or the other partner isn't privy to such facts, no constructive work can be done seeing the couple together.

Lorraine told me that she wasn't a particularly jealous person and that it wasn't like her to be possessive, suspicious, or insecure. But she was having such a terrible time with Ira's friendship that she wanted to insist that they move across the ocean. She also recognized that moving away was a drastic solution and that the triangle she described could occur somewhere else.

Trust Your Gut

It's not useful to drown in emotions or to allow our reactivity to propel us into unwise speech or action. In order to get past a purely reactive voice, Lorraine needed to get a grip on her anxiety and intensity. At the same time, she needed to listen to what her body was telling her. Her jealousy and anger reflected her good instincts rather than off-the-charts possessiveness. Although Ira denied any sexual or romantic attraction between him and Jo, it would have been wildly naive to believe him. He was probably lying either to Lorraine or to himself. Even if his relationship with Jo was in an intense platonic phase, in a matter of time—and with the right opportunity—it could turn into something more. In any case, it was developing at Lorraine's expense.

The dilemma Lorraine faced had no single or obvious solution. Different couples would handle such a situation differently, as might the same couple at different times. Early on in my marriage, I gave my husband my blessing to foster friendships with other women, and to go out with them even when I felt threatened. As I got older, I'd stomp in protest if I felt threatened, as I would to this day. In addition to stomping, I'd be totally open about how vulnerable and threatened I felt. Most couples, even those who place a high value on individual freedom, expect some healthy reserve in their partner's other intimacies.

Lorraine's experience serves as a reminder that sexual temptations are a reality of life. To deny that one's partner (or oneself) is vulnerable to powerful outside attractions can be a form of sleepwalking. No marriage is "affair-proof," and proximity often leads to overinvolvement. Lorraine didn't cause Ira to move toward Jo, nor could she have prevented it. But Lorraine was suffering. She needed to figure out how to express herself with words that might lead to greater clarity and resolution of the problem.

Nurturing the Marriage—and Continuing the Conversation

There were many times when Lorraine felt like saying, "I can't stand this! End your relationship with Jo right now, or I'm filing for divorce!" The problem with such a reactive ultimatum is that it can invite more subterfuge and lying. The unfaithful partner, feeling controlled, moves more toward the third party. Threats do nothing to guarantee fidelity or bolster trust. Nor is it useful to threaten divorce from a position of anger and distance.

A more viable approach is to stay connected and continue the conversation about the problem without distancing or blaming. It was worth taking this difficult route because this relationship had a significant history behind it. Lorraine wanted to do all she could to save her marriage, and failing that, she wanted to end it in a solid way, knowing that she had handled her own part as well as possible. So I encouraged Lorraine to pay attention to her marriage and to try to nurture it. It wouldn't help matters if she distanced or emotionally disconnected from Ira. Rather, Lorraine needed to move toward him with the goal of strengthening their bond by being a loving and generous partner and by letting Ira know how much she needed and loved him. I wasn't suggesting that Lorraine take an accommodating or placating position, but rather that she give this valuable relationship her best shot—and her best self.

At the same time, I helped Lorraine keep the lines of communication open with Ira, so that the subject of his other relationship would not go underground. Attractions kept secret from a partner are more likely to intensify and be acted upon. Lorraine needed to continue to voice her concerns and her pain, and to ask Ira questions about his friendship with Jo and about his experience of their marriage. For example, what did Ira think was going on between the two of them that had made this other relationship so important? What purpose did it serve for him? Did he think something was missing in their marriage? Did he feel loss or grief about the fact that he had no sexual experience prior to marriage?

Lorraine did not back down from putting Ira on the hot seat. When Ira insisted, "Jo and I do not have a sexual relationship!" Lorraine said, "What I'm asking, Ira, is this. If I were a fly on the wall watching you and Jo, would I see anything that would upset me? Would I see something that I might think was occurring at my expense?" Lorraine asked such questions when she felt connected to Ira, not when they were fighting and distant. She held the connection—eye contact and all—as she invited him to tell her the truth, no matter how painful.

Lorraine kept asking Ira for a commitment to honesty, because honesty is the only foundation on which trust can be built. As she continued to voice her vulnerability and engage him in an ongoing conversation, she did the best she could to approach him without being critical, blaming, or controlling. She said, "Ira, it would be devastating to hear the truth from you if you are sexually or romantically involved with Jo in any way. But I don't want you to lie to me or protect me from the truth. Whatever the reality, I'd stay in the marriage long enough to struggle with the issue and to try to get some clarity about it, so we could make the best decision about where to go from there. If you lie, you're putting our marriage in greater jeopardy."

Lorraine's biggest challenge was getting a grip on her own emotions. An extreme response in either direction (whether minimizing or going off the deep end) wouldn't be helpful. When we feel threatened to the core, finding a middle ground is hard, but that's what Lorraine was aiming for. If she could only fight hysterically, or alternatively, if she could only ignore the threat, she would run the risk of driving the whole matter underground.

During this difficult time, Lorraine needed to take good care of herself. To this end, I encouraged her to establish and strengthen connections outside her marriage and to reach out for all the support she needed and deserved. At first Lorraine treated the crisis like a shameful secret. Her silence on the matter only increased her sense of shame, when there was nothing shameful about her situation to begin with. Lorraine needed to talk openly to trusted friends and family because her situation was too difficult to tough out alone or even with therapy as the only resource for input.

I also encouraged Lorraine to engage in creative and healthful activities—anything that would fuel her strength and sense of well-being. She insisted she wasn't even up to loading the dishwasher, but she needed to push herself. Grounding activities—yoga, dancing, walking, attending cultural events—were an essential part of her survival plan. In addition to it being a generally good idea to live healthfully, these activities help lower anxiety and foster a more solid self, which is the foundation for a clearer voice.

Setting a New Bottom Line

Despite Lorraine's efforts to do the right things and stay calm, she felt increasingly crazy. She couldn't concentrate, and she became obsessed with Jo. During a weekend when Ira was out of town, she invaded his privacy by going through his desk and trying to get into his e-mail, in search of clues that something was going on. She hated the person she was becoming and resented the loss of her old self. During one therapy session, she said, "Maybe I'm losing my mind."

I assured Lorraine that she was having a perfectly normal response to her situation. Feeling like she could go crazy was a signal that she needed to bring the conversation to a different level and let Ira know that she could no longer tolerate his other relationship. She needed to set a new bottom line. She agreed, and proceeded to tell him something like the following:

> Ira, I understand that this friendship is very compelling to you. I understand that you're drawn to Jo. But I can't put up any longer with a relationship that feels like it's taking place at my expense. You say that nothing is going on, but I'm not able to stand how important Jo has become to you. I've been feeling crazy, and our marriage is suffering. I need for you to put limits on this relationship and restrict it to work.

Angry fights followed. Ira would say, "You can't control me, I'm an adult. I don't police your relationships, and you're not going to police mine!" Lorraine did not back down. She persisted in voicing her limits and her pain. She told Ira it wasn't her intention to control him, nor did she have the illusion it was possible to do so. He would choose what to do about this relationship or any other. But Lorraine continued to tell him she could no longer live with it. She said, "Ira, I don't want to monitor your relationships. But I thought I was the number-one person in your life. You know how much pain I'm in. So, what does this relationship mean to you that you continue it anyway?" Ira responded with anger and defensiveness, and the emotional climate of their marriage became increasingly strained.

During one therapy session, I asked Lorraine to predict how long she could live this way, if nothing changed. A month? A year? Ten years? Lorraine's response was immediate. "Not a year," she said. Hearing herself say this out loud helped her to voice the ultimate to Ira. She went home and told him that she couldn't live with the current situation for another year, maybe not even for another month. As desperately as she wanted their marriage to work, she was in too much distress. Lorraine had said similar words before, but now her intention was different, and so were her actions. She turned her primary focus toward what she needed to do for herself, if Ira didn't quickly end his relationship with Jo. She stopped focusing on Ira and Jo, and focused relentlessly on her work goals, support network, and plan. For her own immediate survival, she lined up a temporary living situation and began planning her finances in case she moved out.

When people venture into an affair (romantic or sexual), they enter an altered state of consciousness. Sometimes the kinds of conversations Lorraine was trying to have with Ira are enough to solve the problem, but sometimes people awaken only when their partner refuses to allow business to continue as usual—or consults an attorney. Lorraine turned a corner when the option of leaving her marriage became a real possibility to her. By this time, considering separation and divorce was more liberating than terrifying. It helped her to conduct herself in a reasonably calm way as she told Ira that she wanted a separation if he didn't end the relationship.

A Critical Turning Point

When faced with the loss of his marriage, Ira woke up. He became depressed and anxious, and entered therapy himself. In addition, Lorraine and Ira started couples counseling to help them heal and learn from their painful experience. The healing process required Ira to be brutally honest with himself and Lorraine, and to open himself up to the pain caused by his deception. He hadn't had intercourse with Jo, but their relationship had been erotic and romantic, with some "making out," which Lorraine requested he spell out in specific detail. At Lorraine's insistence, Ira transferred to a different office so he and Jo wouldn't be in close proximity. He also wrote Jo a letter that he shared with Lorraine, explaining that he needed to end all contact with her because of his loyalty to his marriage and because he didn't want to do anything to threaten it or hurt Lorraine.

Every new revelation from Ira evoked an avalanche of painful feelings for Lorraine. Sometimes it really hurts to be right. It was a long, slow process before Ira could earn back Lorraine's trust and make her feel loved and chosen. They could not just quickly move on, as Ira frequently hoped, but eventually they were able to recover and restore intimacy as Ira worked hard to earn back Lorraine's trust and to make a commitment to truth-telling and to their marriage. It doesn't always happen this way. The most important point is that Lorraine spoke with clarity and courage throughout this excruciating process. She was honest with herself and with Ira, and she clarified her thoughts and feelings without getting stuck in a position of distance or blame. She gave her marriage the best chance of succeeding, and she tapped into her strength to leave if need be.

Both Clara's and Lorraine's stories illustrate that only when we're clear in our own minds and hearts that we can no longer live with something can the speaking part—although still not easy—become possible. There are no shortcuts to getting there; nor does clarity evolve from thinking and reflection alone. Our gut responses tell us over time what we can and can't live with.

Sometimes we become clear about our position only by wrestling with difficult and uncomfortable feelings that push us to speak or act. The challenge is to sort out our anxiety-driven reactivity (our Dad had affairs, so we go nuts if our partner flirts at a party) from the deeper wisdom of our body. When the pain of the current situation becomes greater than our fear of the unknown, we find the courage to speak differently and to act on our own behalf.

When it comes to unraveling a marriage of many years, clarity is rarely a hundred-percent deal. I've worked with many people who have come to a solid and hard-earned conclusion about the rightness and necessity of leaving a marriage. Yet even as a woman announces her final decision, the voices in her head are saying: "Maybe there's more I can do to make this work." "Maybe he'll change, after all." "Maybe this is a terrible mistake and my kids will hate me forever." What happens over time is that the "maybes" have gotten smaller and weaker, and the resolve to leave has gotten larger. Even when we achieve the clarity and strength to leave a relationship, we can still experience painful ambivalence and self-doubt.

DON'T THREATEN DIVORCE—*DO* TALK ABOUT IT!

To say, "If these things don't change, I'm not sure I can stay in this relationship," is to voice the ultimate bottom line. People threaten to divorce or break up in the heat of anger, which isn't helpful or fair. Nor should you bring up divorce as an attempt to punish, scare, shape up, or shake up the other person. And surely you shouldn't feel compelled to mention divorce simply because it passes through your head now and then. Many married folks entertain fantasies about divorce yet are far from acting on it.

That said, talking about divorce is important if you're thinking seriously about it—even ambivalently. If you're going back and forth about it in your mind, you need to consider sharing your struggle with your partner. If you do eventually terminate the marriage, a partner will be better able to handle a loss that can be anticipated and planned for. Everyone has the right to know just how high the stakes are if they choose to continue to behave as usual. You owe your partner honesty about a matter that so deeply affects both of you.

I've seen any number of devastated men in therapy who tell me their wives left them out of the blue. The wives, however, claim to have voiced their anger and dissatisfaction for a long time. Both are right: *he* hasn't listened well enough; *she* hasn't shared her thoughts about divorce clearly enough or early enough in the process. Often the wife does not make a serious issue of divorce until she's finally made up her mind to leave. Any changes the husband then agrees to make are too little, too late. In the end, neither spouse has had an opportunity to test the potential for change in their marriage.

Exceeding His—or Her—Threshold of Deafness

Divorce is one of many issues that deeply affects a partner's current decision making and future planning. On such matters, we need to make ourselves heard, rather than conclude that the other person can't hear. Sometimes the word *divorce* has been thrown around so much it's become a hollow threat, or your partner just can't imagine that you'd ever leave. When this happens, you need to push the conversation to an entirely different level to be heard.

While it doesn't help to overfocus on divorce, you can't approach the subject in one hit-and-run conversation either. Some people absorb difficult information best in a brief note, so if spoken words aren't getting through, you can also put your thoughts down in writing. For example, a client of mine, Julia, recently wrote a note to her female partner of nineteen years that said:

> I'm writing this note because I don't think you're hearing me. Maybe I've mentioned "divorce" so many times in our fights that it sounds as if I'm just trying to threaten you or shape you up. That's not true anymore. You need to decide whether you want to work on our relationship and make some changes. But I don't think I can continue the way it is. On a 1–10 scale of divorce, I'm about at a 9. If nothing changes, I predict I'll be a 10 within a year.

Here are four reasons why it's necessary to make yourself heard if you're seriously thinking about divorce:

1. *It's not fair to conceal crucial facts that affect your partner.* A spouse has a right to know essential facts in order to think clearly about the present and to plan for the future.

2. *The loss of a spouse or a partner should never come out of the blue.* Many couples constantly fight, blame, complain, and angrily threaten divorce—but never take action. That's why, if you're truly considering leaving, you need to tell your spouse/partner in a different way. You'd expect no less from your boss at work, right? You wouldn't want a supervisor to be criticizing you and giving you negative feedback when the real message is, "These are the specific things that need to change in order for you to keep this job."

3. *A partner will have the best chance of deciding whether to make the necessary changes (go to marital therapy, find a job, become a partner in housework and parenting) if he knows that the problems are so serious that you're contemplating divorce.* If you're very clear that you can't continue with the status quo, your partner will also be clear about his willingness (or unwillingness) to change and about how much the marriage means to him.

4. *Talking frankly about divorce will make the possibility of divorce more real to you.* Doing so will ultimately add greater clarity to your own thinking, whatever path of action you choose.

A couple of caveats: Obviously, we should never mention divorce (or anything else for that matter) if there is any possibility that a partner will become violent or out of control. In such a case, we first need to seek appropriate help and ensure our physical safety. Nor is it wise to begin a serious talk about divorce if we suspect that a partner might do something sneaky with money that would jeopardize a fair and equitable financial settlement. In such a case, it's wise to first consult an attorney. Finally, if you've already made up your mind to leave, it's not fair to involve your partner in conversations that imply you're still willing to work on the marriage.

MOVE TOWARD YOUR PARTNER WITH LOVE

If you're trying to set limits, take a firm position, or voice the ultimate in your relationship, you may not feel like moving toward your partner in a loving and respectful way. But if you can't warm your partner's heart, you won't get anywhere crossing your arms over your chest and making all kinds of heavy "I statements" about what you won't tolerate.

If your motivation is to give your relationship the best chance of succeeding—or simply to be heard—the first order of business is captured in the advice I gave Lorraine when she was struggling with Ira's other relationship: You need to pay attention to your relationship and nurture it, you need to move toward your partner in a loving and generous way, and you need to avoid distancing or disconnecting emotionally. You can do this—and also clarify your wants, expectations, limits, and bottom line.

Whether the subject is divorce or the dirty dishes, you don't need to voice your bottom line in somber, heavy, or angry tones. Clarity of voice comes from inner conviction about oneself and from deeply held values and self-respect. To be heard, you may need to lower the volume and intensity, rather than raise it.

When the emotional climate of a marriage is predominantly negative, you have some groundwork to do before clarifying a bottom line. No one will listen or speak well if the overriding sentiment of the relationship is hostile, critical, or distant. Nor will you get anywhere taking a position about a hot issue if you're coming from a place of angry reactivity, self-righteousness, or criticism. Instead, you need to be a model of the kind of behavior you want from your partner, not a critic. You need to treat your partner the way you'd like to feel about him, in order to evoke the positive feelings you've lost touch with.

Taking such positive action is incredibly difficult when a relationship has become a source of chronic anger or dissatisfaction. But before you rush in to clarify your bottom line or to find a divorce attorney, it's a good idea to first test out what's possible by giving voice to the positive.

CHAPTER 10

Warming Things Up

It's obvious to people that having an authentic voice requires us to take positions and speak to our dissatisfaction. What is less obvious (but no less difficult) is that having an authentic voice requires us to speak to the positive. If you can't find much positive to speak to in the other person (whether it's your mother or your partner), you've lost perspective. Every individual has some strength and goodness. Every relationship has some positives, even if both partners have forgotten how to notice and comment on them.

The longer you're with someone, the more vulnerable you are to selective attention. You automatically register and give voice to what bothers you, and you automatically fail to give praise and to voice your appreciation. Many folks welcome constructive criticism during the courtship phase of a relationship, but tolerate it less well over time. Most importantly, nobody values criticism if there's not a surrounding climate of love and respect.

If you can warm up the emotional climate, or at least put a few cracks in the ice, you'll be better able to speak your own truths—including your complaints—and maximize your chances of being heard. Only then can you really begin to determine what's possible in a relationship, and only then can you nourish your own soul by giving voice to the positive.

CONSIDER THE EMOTIONAL CLIMATE

When the emotional climate of a relationship is spontaneous and relaxed, there's a lot you can get away with. Just within the past twenty-four hours, I've said to my husband, Steve:

"That's a nerdy shirt. Wear the other one."

"You didn't call your cousin back? That's terrible! Make sure to call her today."

"Why didn't you tell Ben's friend that he can't call the house at one-thirty in the morning? You acted like it was fine that he woke us up."

When Steve is in a good mood, I don't consider how I say things. If I'm intense or obnoxious, he'll ignore my tone or tease me out of it. He'll accept what he finds useful in my complaints and step aside from my negative judgments. I love the fact that Steve and I feel free to criticize each other, because that's one way we learn from each other. Learning from each other has always been a large part of our mutual attraction.

But when things aren't warm between us, Steve is definitely not up to hearing my complaints. He may react to the slightest negative edge in my voice. When we're both high in reactivity, tempers can flare. "You're not going to talk to me *that way!*" Steve insists. "I don't appreciate the fact that you're acting like we can just move on to another subject, when you haven't apologized for being rude to me." In response, I may argue that I wasn't as rude as Steve is making me out to be, that he's totally overreacting and blowing things out of proportion, and that he's the one who should apologize. He argues that I need to take responsibility for my behavior and apologize first, or nothing will move forward.

Usually it doesn't take long before one of us changes course and does the necessary repair work. Steve may knock on the door of my study and say, "Please forgive me. I acted like a jerk. Give me a hug." He has invented a very silly "1-2-3-let-it-go" ritual that makes me laugh, so I can't stay angry. Or I'll take the initiative to apologize: "We're both under a lot of stress. I apologize for my part. Let's just drop this and try to get along." I may be convinced of my relative innocence and believe "my part" is only 2 percent, but I don't feel compelled to make this point. These days, we're both pretty good at repairing a disconnection before much time has passed. We can also let each other know up front if we're not in the mood to hear anything negative.

Some very fine people can't tolerate much criticism, even from—or *especially* from—their partner. When this is so, we need to consider how to choose our words and consider how to put things, even if we feel compelled to spontaneously say whatever we think. If we're not being heard, more of the same won't help. Plus, being real is hardly a virtue when we're having a negative impact on the relationship. And there are many different and equally real ways to make a point.

He Won't Listen/She's a Bitch

A greeting card says, "If a man is alone in the forest with no woman there to criticize him, is he still a schmuck?" It's a takeoff on the popular Zen koan asking whether a tree falling in the forest makes a noise when there's no one there to hear it.

Men roar with laughter when I share this card at lectures, and many comment on how perfectly the card speaks to their experience. Women, however, tend to have a muted or negative response. "If these poor guys feel so beleaguered by criticism," one woman said, "why don't they shape up?" She went on to share how angry she feels when she has a legitimate complaint but is ignored by the very person who is supposed to be her loyal advocate and most reliable support. "Yes, I feel like a nagging bitch," she said, "but if he wants me to stop nagging, he can pay attention to what I'm asking him to do."

It's easy to appreciate both perspectives. We do feel pain when we're the target of constant criticism, just as we feel pain when our words fail to move our partner and our legitimate complaints are ignored. It's easy for two people to reach an impasse where closeness, conversation, and collaboration have broken down.

When we're feeling angry, it's hard to take positive new steps to speak differently ourselves. We're convinced that the solution is for the other person to change. *She* insists that the only way the marriage will improve is for him to become more responsible and pay more attention to the kids. *He* insists that she needs to become less critical and controlling and more appreciative of all he does for the family. When this happens, nothing will change until one partner makes an effort to calm down, or better yet, to warm things up.

WHAT'S GOOD ABOUT BEING NICE?

Family therapist Monica McGoldrick was conducting a workshop called "Marriage, Divorce and Re-marriage," at the Menninger Clinic. Someone in the audience asked what advice she would give to a woman who couldn't get clear about whether to stay in an unhappy marriage or leave. "I'd tell her to do an experiment for nine months," McGoldrick replied. "During this time, she should make a 100 percent commitment to being the very best partner she can be. She should be the partner that she wants her spouse to be for her. She should also work on herself, including her relationships with her family of origin. If she's still thinking about divorce after that, she probably should go ahead."

A colleague sitting next to me bristled at McGoldrick's advice. "She's a feminist?" she whispered to me. "And her solution is for the woman to become the perfect wife!" When we talked later, I learned that my colleague had heard McGoldrick prescribing dishonesty. The woman contemplating divorce was surely feeling angry and critical, and undoubtedly had many legitimate complaints that shouldn't be suppressed or glossed over.

Actually, it's a very rare person who can sustain the motivation to persevere with this difficult experiment for several weeks, let alone for nine months. But unlike my colleague, I agree with Monica McGoldrick's advice and the philosophy behind it—which is not about tolerating unacceptable treatment. Sometimes we have to deliberately refrain from excessive criticism and negativity and, instead, experiment with such virtues as kindness and generosity of spirit. We have to stop waiting for the other person to change first. If you're feeling madder than hell right now, the thought of warming up the relationship may feel impossible. Actually, it's not impossible. It's just extremely difficult.

Let's consider a typical couple where neither person found a different voice to create positive change. I'm not presenting this gloomy case history to dampen your day. But I do want to wake you up to the reality of how high the stakes are when you get entrenched in the negative and dig your heels in because you feel like the wronged party.

CRITICISM AND STONEWALLING: A RECIPE FOR DIVORCE

I saw Augusta and Harry for seven sessions of marital therapy before they quit coming; they divorced a year later. Harry said that Augusta constantly supervised him ("Don't put so much water in the pot for the pasta," "Don't use that knife to cut the vegetables") and failed to give him space to do things his way. Harry felt worn down by her criticisms.

I noticed that when Augusta addressed the important issues, she didn't limit her complaints to Harry's *behavior*, which would be fair enough ("I'm angry because you said you'd call when you're going to be late for dinner. This is the second time this week that you didn't call"). Rather she criticized his character ("You don't think about anyone but yourself. I can't count on you to do anything you promise you're going to do"). She engaged in all the below-the-belt tactics that books on communication wisely teach you to avoid. She globalized, moralized, lectured, interpreted, and all the rest. Intermittently, she retreated into a cold and bitter distance. Through pursed lips, she would quietly say, "Nothing," whenever Harry asked her what was wrong.

If you were a fly on the wall listening to a session or two of this couple's marital therapy, you might think, "that poor guy," or "what a bitch." Neither perception would be the slightest bit objective or accurate. These were both loving and good people. Like many modern couples, they did fine until their first baby came along and swept them up in the strong current of traditional gender roles. He rolled up his sleeves to earn, earn, earn, while she took over at home. By the birth of their second child, all the circuits were overloaded. Augusta, who also worked outside the home three days a week, was exhausted and depleted. She was the one noticing and doing about 80 percent of the child care and housework. She had lost her earning power, while Harry had increased his. She was also the one who had become uninterested in sex, as the unequal distribution of domestic tasks took its physical and emotional toll.

These are not minor problems that go away when ignored.

Many modern couples are convinced that they won't fall prey to the old roles, but it's hard to swim against the cultural tide. Augusta, who was in the most pain about everything except the sexual problem, wasn't addressing her own needs directly, with an eye toward renegotiating the marital contract in a fair-minded way. Like many women, she felt that she should be able to handle it, even as she complained that the "it" was unfair and too much. Keeping the family running smoothly is, after all, what women are "supposed" to do. If we don't do it happily or well, we have no one to blame but ourselves. Like many women, Augusta disqualified her legitimate anger, only to assume a generally blaming attitude toward Harry.

Self-Focus vs. Other-Focus

Beneath Augusta's critical focus on Harry lay her lack of clarity about her own priorities and whether she was entitled to claim them. She wanted Harry to be home more, but she didn't feel empowered to ask him to cut down his hours at work because his job paid so much more than hers. She learned what the culture teaches—that the man's career is sacred and can't be tampered with when kids come along, and that mothers alone will make the enormous changes required by children. Augusta expressed bitterness that she alone was left struggling with the balancing and juggling dilemma, but then she'd do a turnabout and disqualify her own anger, saying that no one was *making* her stay home, and she wasn't sure she wanted to work more hours anyway.

Harry and Augusta were so reactive to each other that they almost never talked calmly together. They fought (or he withdrew), but they didn't sit down together, examine the facts, and work together as partners to find new options for spending time differently. When Harry said, "I have to stay late at work to get the job done," the conversation stopped there. They never crunched the numbers to see how their family might survive financially with a new plan, nor did they see the possibility of other options. For example, could they cut down on certain expenses so that Harry could work fewer hours and be home by dinner? Would Augusta prefer to work more, even though her earning power was less than Harry's? Were there creative solutions around the question of "Who does what?" that might give each of them more of what they wanted? What *did* each of them really want, anyway?

Augusta and Harry never became self-focused enough to examine their own core values and beliefs. Did Harry, for example, really want to follow in his dad's footsteps and develop his career at the expense of knowing his family? What sort of father and husband did he want to be? Getting clear about what really matters is almost impossible when one's energy is dissipated in reacting to one's partner. It wasn't that Harry and Augusta weren't creative or smart enough to think through these questions. But they were both so locked into unproductive patterns that they didn't give themselves a chance to even ask these questions, let alone answer them.

Harry, for his part, was withdrawn and defensive. In withdrawing, he stopped moving toward his wife lovingly and affirmingly. In his defensiveness, he listened for the part of her criticism that he did *not* agree with, rather than listening for the part he *might* agree with. He didn't pay attention with an eye to really getting it, but rather to refuting what was exaggerated or inaccurate. He was understandably bitter that his hard efforts at being the breadwinner seemed to go unappreciated. He felt as if so much was expected of him in the two different worlds of work and family that he could never satisfy everyone. Feeling beleaguered and unacknowledged, he shut down.

Most importantly, Harry felt unable to tell Augusta that her criticisms bothered him, that he couldn't tolerate feeling constantly scrutinized and evaluated. He never found his voice to say, "I need you to find a different way to talk to me about your important concerns, because I'm feeling flooded by so much negativity. I'll listen to you better if you approach me with respect." Instead of clarifying this position and standing behind it as long as necessary, Harry allowed Augusta to go on and on. He then ignored her complaints, for example, by "forgetting" to call when he was delayed at work. He wouldn't look up from the newspaper or television when Augusta confronted him. When Augusta accused Harry of being passive-aggressive, he shut down further. Augusta's own self-esteem plummeted as she found herself becoming the stereotypical nagging wife.

Is It Honesty or Reactivity?

Augusta viewed herself as open and honest with Harry, because she equated honest communication with the uncensored expression of thoughts and feelings. It's indeed wonderful to have moments in a relationship when we can spontaneously share everything without thinking about it. But at the very moment Augusta would lay claim to "being herself," she was anxiously reacting to Harry. In turn, he was also anxiously reacting to her. Silence and stonewalling are simply ways of managing intense emotional reactivity.

Any therapist who works with couples (including, say, a mother and a daughter) sees a lot of colorful and intense interactions where each party blames the other for the problem. It may appear that authentic feelings are being voiced, but there is little exchange of anything real, except perhaps for the palpable presence of real pain. Such momentary "honesty" shuts down the lines of communication rather than widening the possibilities for truth-telling.

When the emotional climate is intense, a couple may behave like two nervous systems hooked together. Neither party can identify and calmly address the important issues, listen to the other objectively, or take a position without blaming or telling the other what to do. The contagious reactivity between two people can be so high that almost any topic triggers immediate intensity. Within moments both persons are rigidly polarized in opposing camps, unable to consider any viewpoint except their own.

The freedom to give full vent to our feelings can attest to the durability of a relationship, and fighting is certainly one way for people to stay connected and let off steam. In my own marriage, Steve and I have had good fights and bad (including *very* bad) fights. Our marriage is definitely more intimate because we *can* fight. When we fight well, we emerge from the fray with a deeper knowledge of ourselves and the other. When I work with couples who tell me they never fight, I ask them, "Why not? What's wrong?"

But when two people are stuck in reactivity, it's a different story. We're not in a process of knowing the other person and being known, of refining and deepening our disclosures to the other person. To move in this positive direction, at least one person has to begin to warm up the emotional climate, to be a more loving presence, or if not that, at least a calmer one.

"We Love Each Other, But . . ."

When I work with couples like Augusta and Harry, I often recommend a book by couples therapist Ellen Wachtel called *"We Love Each Other, But . . ."* (a book Steve has affectionately retitled *"We Love Each Other's Butt"*). This is the wisest, easiest-to-follow book I know for helping couples warm up a relationship when one or both parties are locked into the negative. It also includes specific guidelines for handling explosive arguments and breaking vicious cycles.

If the couple is up for tackling a more difficult read, I suggest *Love, Honor and Negotiate* by family therapist Betty Carter. It's the only book I know that looks squarely at gender inequality in marriage and contains great advice for making marriage work. It's far more difficult to put into practice, because it challenges our gendered assumptions about what men and women feel entitled to and responsible for once kids enter the picture. These assumptions establish outposts in our head even as we fight against them.

Not everyone loves my reading list. Harry started to read *We Love Each Other, But . . .* and never finished the first chapter because "he didn't have time." Augusta got through about sixty pages and let me have it. "You wrote *The Dance of Anger!*" she said, "and you tell me to read *this* book!" She didn't relate. "I'm supposed to establish a loving atmosphere by conveying my admiration and being the president of his fan club! I'm supposed to warm his heart and do little things to make him feel special! You give me a book that teaches me to be *nice!* Give me a break! I've spent most of my life being nice, and I'm not going back there."

Augusta's words resonated with my colleague's response to Monica McGoldrick's prescription at the conference. She insisted on doing what felt "real and natural," which actually meant doing what was reflexive and familiar—a life on automatic pilot in a marriage on a downward course. Augusta simply wasn't able to step aside from her own intensity long enough to consider changing course. Neither was Harry.

Why Be Nice?

Why should we practice kindness when the other person is behaving badly? Such advice may seem incongruent with the challenge of speaking our truths, bringing up the hard stuff, defining our differences, and clarifying a bottom line. Actually, kindness and generosity of spirit lay the groundwork for each of these things.

Of course, Augusta's sentiments are perfectly understandable. The injunction to be nice has served to paralyze women and hold us in place. Pressures to please and protect others can limit our creativity, vitality, and imagination, silence our legitimate anger and protest, and keep us close to home. In the name of being nice, we can make a commitment to security, sameness, and safety, rather than to truth, courage, and honor. We learn to be the soothers, protectors, and steadiers of rocked boats—when we'd be better off using our voice to make waves. We may tolerate rude or disrespectful behavior in a partner. This is not the kind of "nice" I suggest we aim for.

But kindness, timing, and tact are not the opposite of honesty: Rather they are precisely what make honesty possible with the most difficult people and in the most difficult circumstances. There is no virtue in speaking to others in a way that makes it impossible for them to hear what you have to say or to appreciate the truth of your position.

Perhaps, like Augusta, you're saying to yourself: "There's no way I'm going to put in the effort to warm up this relationship when my partner is doing X, Y, and Z! I don't *feel* like being emotionally generous." Well, if this is the case, you can *pretend*. You can treat your partner the way you'd *like* to feel about him, in order to evoke the positive feelings you've lost touch with. You can behave like the partner you want him to be for you. You can be a model, not a critic. Doing this for even one week can yield valuable information.

ANGER: A DOUBLE-EDGED SWORD

My point is not that you should deny your anger or ignore its sources. On the contrary, anger is an important signal that something is wrong. It always deserves our attention and respect. Anger can sharpen our passion and clarity and inspire us to speak honestly and truly. It can motivate us to say no to the demands and expectations of others, and yes to the dictates of our inner self. Our anger can help us clarify where we stand, what we believe, and what we will and won't do. Our anger tells us when the other person has crossed a line that shouldn't be crossed. In all these ways, our anger preserves the very dignity and integrity of our voice. If we didn't have our anger to motivate us, our fear might lock us into passivity, silence, and accommodation.

But the opposite occurs as well, perhaps even more frequently. When we angrily confront another person, convinced that truth is on our side, we may move the situation from bad to worse. Anger can create tunnel vision that leads to a narrow, fixed view of what is true and whose truth counts. When tempers flare, the capacity for empathy, listening, and creative problem-solving that considers the needs of all is diminished. It may become harder for two people to uncover and share their own truths, to hear each other, or simply to stay in the same room. Everything becomes a bigger deal than need be, or we stop talking about a subject entirely.

Augusta, for example, did have something to be angry about. Too much responsibility was falling on her. But her complaints were either not clearly voiced or put forth in a way that elicited Harry's defensiveness rather than his empathy. Augusta was quick to get angry, but when getting angry got nowhere, she kept doing more of the same instead of considering creative options for approaching Harry in a new way. Perhaps nothing Augusta could do or say would get Harry's attention or elicit his empathy. It's abundantly clear that we can't control other people's reactions. But the choice to not change ourselves is a surefire way to keep things in the same place or move the relationship from bad to worse.

Sometimes a genuine show of fury will get the other person's attention, but not if it's habitual. Expressions of anger and criticism are simply not constructive if we've gotten out of the habit of observing what's going right and commenting on it, or if we complain but don't really challenge the status quo.

A WARNING ABOUT NEGATIVITY

When couples like Augusta and Harry are stuck in negativity, I may try to boot them off automatic pilot by telling them about John

Gottman's research findings. After studying thousands of married couples, Gottman concluded that certain kinds of negativity, when allowed to run rampant, are lethal to a relationship. He calls them "the four horsemen of the apocalypse," which clip-clop into the heart of marriage and destroy it. In his book *The Seven Principles for Making Marriage Work,* Gottman describes the four horsemen in the order they're likely to appear and wreak havoc. Here's an abbreviated summary:

HORSEMAN 1: CRITICISM

A "criticism" is a personal attack that includes some negative words about your mate's character or personality. For example, "Why do you keep putting your friends ahead of me? I always come last on your list. We were supposed to have dinner tonight." This sort of criticism differs from a constructive complaint that addresses a specific action or behavior at which a spouse has failed. ("You were supposed to check with me before inviting anyone to dinner. I wanted to spend time alone with you tonight.") To turn a constructive complaint into a corrosive criticism, Gottman suggests adding the gibe: "What's wrong with you?"

HORSEMAN 2: CONTEMPT

Contempt can be conveyed in many forms, including name-calling, sneering, eye-rolling, mockery, hostile humor, sarcasm, cynicism—any nasty or mean-spirited attempt to put the other person down. For example, she complains that he's late for dinner, and he says, "What are you going to do, sue me?"

HORSEMAN 3: DEFENSIVENESS

Defensiveness is a way of saying, "The problem isn't me, it's you." We fail to really listen, to consider our part in a problem, and to apologize and change our behavior. When our partner complains, we argue, attack back, bring up his or her faults, and climb further up on our high moral ground.

HORSEMAN 4: STONEWALLING

Stonewalling occurs when one partner tunes out the other and disengages from the relationship. We turn away, sit there like an impassive stone wall, leave the room, or somehow communicate that we couldn't care less what the other person says or does. We won't let our partner's words influence or affect us. Gottman states that people stonewall as a protection against feeling flooded, and that men stonewall more than women.

Obviously, one finds all four horsemen in the best of relationships, but when they take up permanent residence—and when there is *a failure of repair attempts*—Gottman claims he can predict divorce with an accuracy rate that reaches well above 90 percent.

REPAIR ATTEMPTS

Gottman's four horsemen don't mean that happy couples don't fight. On the contrary, he reports that happy couples have loud arguments—even screaming matches—without necessarily harming their marriage. Nor do these successful couples necessarily do anything that resembles "active listening" or other good communication strategies when they're upset. What does matter, according to Gottman, is a deep friendship, meaning mutual respect and enjoyment of each other's company—which leads to successful repair and reconciliation.

What is a *repair attempt?* Gottman uses the term to refer "to any statement or action—silly or otherwise—that prevents negativity from escalating out of control." He gives the example of a couple fighting about whether to buy a Jeep or a minivan, with the conflict escalating into the higher decibels. Suddenly the wife sticks her tongue out in perfect imitation of their four-year-old son, and the husband, anticipating she's about to do this, sticks his tongue out first. The tension is defused as they both start laughing.

My husband, Steve, apologizes easily, and he often relies on humor because he knows I don't stay angry in the face of his irrepressible silliness. Some people can't apologize, particularly if their partner insists on it, but there are many other ways to deescalate tension and make reparative efforts. One person may say, "Let's take a break. I need to calm down. Let's each think about what we're both saying, and we'll talk about it after dinner." Or even, "Hey, don't yell at me, I won't stand for it." The failure to initiate repair attempts—*or the failure to respond to a partner's attempts to do so*—is a flashing red light in any relationship. When we refuse to take our partner up on her repair attempts, or we don't give a partner a fair way out of a fight or tense conversation, we need to ask ourselves if we are truly more invested in our anger and dissatisfaction than we are in changing the tone of our relationship.

One final point. Gottman also claims that if a couple can keep up a five-to-one ratio of positive to negative statements or interactions, then the four horsemen aren't lethal. The idea of making a deliberate effort to get our communications in line with this ratio may sound a bit absurd. But it can be an interesting experiment, because there's nothing real or authentic about having our brain stuck in a negative groove.

STOP, IN THE NAME OF LOVE

To borrow a line from Diana Ross and the Supremes, couples can learn to "stop, in the name of love." If relying on our own wits and humor isn't working, couples can establish a "STOP rule"—a "time-out." Doing so is especially important when one person's anger is spilling over, or one person feels diminished or unsafe.

Couples need to establish some ground rules for fighting rather than assuming that feeling enraged ("I can't help myself") gives you license to say or do anything. If you can't maintain control of your own voice, you need professional help. As psychologist Marty Klein points out, even war has its rules. In World War II, for example, there was a rule that you couldn't bomb the enemy's hospitals. Surely couples can agree on a few rules ("No name calling, no screaming, no bringing up past grievances and hurts during a fight") or can get help to do so.

When I work with couples where one or both partners say they can't control what comes out of their mouths, I sometimes suggest they follow Ellen Wachtel's ground rules for calling a halt to any argument or interaction that escalates: Both parties need to agree that when either one feels uncomfortable with an interaction, that person can call an immediate halt to the conversation. Both keep a written copy of the STOP rule on an index card in a place like the dresser mirror where they'll see it every day. The rule should say, "We each agree to respect the other's wish to stop the discussion, even if one of us does not think it is an escalating argument." Then, if either party starts thinking, "Oh, no, here we go again!" that person needs to stop the argument in midstream by saying, "This conversation doesn't feel good to me. I'm invoking the STOP rule." During the cooling-off period, each partner must honestly agree to consider the other's point of view. If an issue needs to be revisited, it helps to let at least twenty-four hours pass so that the problem can be approached more calmly and productively.

My point here is not to provide how-to tips on fighting fair but to underscore that all couples need to establish rules about fighting. I also want to encourage you to think about whether you want to use the power of your words to create more distance or intensity, or to diminish, shame, deflate, or put down your partner. The latter may, indeed, be your intention ("My goal is to hurt my partner and make her feel as miserable as she makes me"). But if not, you need to stop your part in escalating fights and ask some difficult questions: What is happening to me in these angry encounters? Am I communicating in a way that is rooted in my values and in line with what I want to achieve in my relationship? Raw, unbridled emotional exchanges may occur in the best of marriages, but in these they are the exception, not the rule, and they are quickly repaired.

WARMING THE OTHER PERSON'S HEART

The opening premise of *We Love Each Other, But . . .* is this: We all know what warms our partner's heart if we stop to think about it, but when people are angry and hurt, they gradually stop doing whatever makes their partner feel warmly toward them. Wachtel points out that you don't have to feel great about the other person to begin to do the things that warm her heart. But you do need goodwill and a desire for a loving relationship.

If that sounds simple, keep in mind that it's usually the simple things we "forget" to voice. The longer couples are together, the easier they fall out of the habit of doing the obvious to create positive, loving, and affirming interactions. At the start of a relationship, we may tell the other person what we value and appreciate, then do the opposite as time goes on.

Wachtel's book sparked an insight about a problem in my own marriage. Steve does many loving and heartwarming things. He brings me cappuccino in bed every morning, which requires him to walk up and down two flights of stairs. He's there to fix my computer or whatever else needs fixing. He often tells me how much he loves me and how lucky he feels to be married to me. But when I read Wachtel's book, I realized that he had totally lost the habit of telling me the *specific* things that he notices and admires about me, something we did a lot with each other when we first got together.

When I thought about it, I couldn't remember the last time Steve said something like, "You handled that situation with the kids really beautifully"; or, "I appreciate that you go on the road so much, because the income you bring in has been great for the family"; or, "Thank you for bringing such beautiful art into the house"; or even, "What a great salad!" It took me by surprise to suddenly admit to myself that I even wanted this kind of feedback. After all, I know Steve loves and respects me. Plus, there is a widespread belief that if you have good self-esteem, you don't need affirmation from the *outside* (this is brazenly untrue, by the way) because you have it from within.

I also realized that I wasn't making these positive comments to Steve, either—not that he was complaining. Interestingly, adults understand that children need us to praise specific traits, qualities, and behaviors. We don't just say "You're a great kid" and "I love you so much." We also say "You did a good job setting the table," or "I think you were very courageous to tell your friend how you felt." Grown-ups also need to hear the specifics, and frequently at that. Surely it's good as well for our own soul to voice them.

After having this flash of insight, my first impulse was to ask Steve to read Ellen Wachtel's book, so he would be inspired to change his behavior. Instead, I took the high road (being a model, not a critic) and began to talk to Steve the way I wanted him to talk to me. I started noticing and praising him for the specific things I had stopped noticing, or simply took for granted after thirty years of living together. I did this for several months before telling him that I missed hearing the specifics from him, and would he make the effort, which he did. I realize that I'm not talking about a huge problem here to begin with, but little things can make a big difference.

Even if Steve never behaved in kind (I still remind him to do so), I would have continued the change I initiated. My "experiment" taught me that noticing the positive and expressing appreciation were actions congruent with my values about the sort of person I want to be in the world. Finding our own voice requires us to examine our core values about how we want to navigate relationships, and not simply to behave *in reaction* to how the other person treats us ("Well, if my son never calls me, I'm not going to call him"). I don't want to enhance and affirm Steve just to make it more likely that he'll respond to my criticisms or requests for change. More importantly, I want to be a person who balances my automatic critical responses with more positive ones. The positive creates the positive, so the more I choose to express appreciation of Steve, the more I actually feel it.

You're the Expert

A man sat in my office the other day telling me that he had tried everything and was at a total loss about how to make any improvements in his marriage. Of course, I didn't believe him. I asked him to come up with three specific actions he could take to improve the situation at home and warm his wife's heart. Immediately, he said, "Well, I could cook her favorite dinner and have it ready by candlelight when she gets home from work. And I could insist on taking the kids shopping next Saturday so she can have a day off to rest. And I could ask her how she's doing with her mother instead of telling her I'm sick of hearing about it." These were great ideas—small and specific enough for him to accomplish quite easily if he genuinely wanted a more loving relationship.

Magazines and self-help books are filled with specific tips on how to make a partner feel valued and special. You don't need this advice. No matter how distant a relationship has become, and no matter how dense you insist you are, there is no expert who knows what warms your partner's heart the way you do. It's getting started and sticking with it that's the hard part. We all tend to focus on what the other person is doing *to* us, or not doing *for* us.

If the overriding sentiment in your conversations with a partner is less than positive, here's your homework: For two weeks, increase your ratio of positive to negative statements to five to one. Be sure to initiate and respond to repair attempts. Experiment with communicating interest, generosity, and love in nonverbal ways, as well as with words and language. Do the small, specific things that warm your partner's heart. Even small steps in this direction will allow you to know yourself and your partner better, a worthwhile venture whether your partner eventually responds in kind or not.

The goal is not to put a patina of false brightness over problems. Rather, you can aim to speak in a balanced way to both the good and the bad. You can use words and actions to create an emotional climate in which people can be open and thoughtful, feel respected, appreciated, and heard, be more of their best selves, and give the relationship the best chance of succeeding. Constructive criticism and loving warmth each prepare the way for the other.

CHAPTER 11

Silent Men/Angry Women

Myth has it that men are more assertive than women, but this is not the case in intimate relationships. Rather, "He won't talk about it" is a commonly heard complaint of women. In the bad old days (that is, before feminism) marital therapists came up with the following strategic solution to the "He won't talk"/"She won't have sex" impasse. The wife was given tokens to dole out to her husband in exchange for, say, twenty minutes of conversation. After the husband collected a certain number of tokens, he could exchange them for "a good schtupping" (Yiddish for intercourse), as a colleague puts it. She would get her conversational needs met, he would get his sexual needs met, and all would be well with their world. Thankfully, this "therapeutic solution" has been discarded. But the woman's complaint still has a familiar ring.

★   ★   ★

It's said that men remain silent because they want to exert power over women, but I don't believe this is so. Men don't sit around in smoky rooms with maps and pins figuring out whether the best strategy for controlling women is through words or silence. Men choose not to talk because they don't know how to make themselves heard, because they believe that problems get worse when you talk about them, because they dread conflict and criticism, or because they fear getting trapped in a conversation that feels awful. Many men can't comprehend the reality of male dominance because they feel anything but dominant in the arena of private conversations.

Thinking in Threes

Men suffer the most profound loss of voice in *triangles,* as my father did when he felt caught between the competing demands of his wife and his mother and when he couldn't take a clear position with either. Every family has triangles that span generations and households. Indeed, we all operate in triangles where conflict in one relationship will inevitably detour into another, making it difficult to sort out what problem belongs to which relationship and who needs to speak up to whom.

The best way to explain how triangles operate is to describe one. Here's a snapshot of one man's impressive move toward finding his voice in an extremely intense yet quite ordinary stepfamily triangle. Even if you've never been part of a stepfamily, the lessons will apply to other relationships you are in.

CAUGHT IN THE MIDDLE

Bill came to see me in a state of great distress after his recent marriage to Alice, his second wife. Everything had gone fine during their courtship, but now the tension in their household was unbearable. Alice constantly complained that Bill's sixteen-year-old daughter, Donna, was unkempt, lazy, and unruly. Alice blamed Donna's mother, Carol, for doing a poor job of socializing her, and she set about establishing and enforcing "new house rules" ("Donna must make her bed in the morning and start attending more to her personal hygiene!").

Alice's abrupt way of taking charge reflected her high anxiety and proved unhelpful. She was negatively riveted on Bill's ex-wife as being "only trouble." Alice not only disliked Carol but also felt threatened by her, so she was unable to support Bill's efforts to co-parent with his ex-wife. Alice was also trying to shape up Donna according to her own values and ideas about how teenagers should be raised. As a result, the relationship between stepdaughter and stepmother went from bad to worse.

This was the picture I got from meeting with Bill—a perfect portrait of the wicked stepmother and the impossible teenage daughter. Let me be clear that neither Alice nor Donna was "the problem" in this family. No one person creates the pattern in which all the other family members participate. Alice simply managed her anxiety differently from how Bill managed his. She tried to shape up other family members by focusing on them in a blaming or critical way, while Bill was a champion distancer. Bill's inability to take a clear position with either Donna or her mother totally overloaded Alice's relationship with both women, and pulled her right into the "wicked stepmother" role.

Stepmothers labor under expectations that even a saint couldn't live up to. When Bill remarried, he expected Alice to take charge of his daughter because, after all, that's what a wife "should do." Bill himself had never even formulated clear beliefs about how to rear

Donna, and he had always deferred to his first wife on parenting matters. Alice jumped in to fill the gap partly because Bill left a resoundingly empty space that begged to be filled.

A Dread of Conflict

Although Bill disapproved of Alice's behavior, he greatly feared expressing his differences. He had felt devastated when Carol left him, and he didn't want a repeat performance with his new wife. In his desperate desire to make his new marriage work, he tiptoed on eggshells with Alice. He watched his adolescent daughter, Donna, act out in response to Alice's attempts to "get her under control," but he himself lacked the confidence to discipline his own daughter. Both Bill and Alice were swept up in the strong current of societal gender roles—he would earn, while she would become an instant mother (just add kids and stir). The arrangement was unworkable.

When I first saw Bill, every family relationship was strained to the breaking point. He was at a loss about how to take a position with the three key females in his life, who then played out the unresolved emotional intensity in their relationships. This is a common way for triangles to operate. Of course, if you're on the outside looking in, it's much easier to feel sympathetic with the distancer, who seems to be a nice guy and terribly beleaguered, than with the complainer, who happens to be in the impossible stepmother role and feeling desperate.

To Bill's credit, he let go of blaming Alice and began bringing more of himself and his voice into family relationships. He was motivated to do this difficult work by his suffering and by his fear that his new marriage might fail. Often people find the courage to speak differently only when the pain of the old way becomes more than they can stand.

Warming His Wife's Heart

Bill learned to speak in new ways he had never imagined. First, Bill needed to warm things up with Alice, to help her feel loved, safe, respected, and central in their marriage. He was totally committed to making their marriage work, but in their tension-filled household, he neglected to do the very things he knew warmed Alice's heart—like cooking her favorite breakfast or arranging a movie date. He lost sight of all the qualities in Alice that drew them together in the first place, and he stopped commenting on them. He no longer conveyed his love and admiration, or made small day-to-day gestures to let her know she was valued and special.

Bill had trouble focusing on the positive when he felt as if he and his daughter were the primary source of unhappiness in Alice's life. But he rose to the occasion, letting Alice know that she was loved and that, despite their family conflicts, she brought joy to his life. He began to make sure they had special time alone together.

Another way Bill let Alice know she was loved and chosen was by creating a clear boundary around their marriage. To this end, he needed to set appropriate limits with Carol. For example, Bill told me when he started therapy that he "couldn't get Carol off the phone" when she called to discuss their daughter. His failure to do so was one of many examples where his silence and passivity occurred at Alice's expense. So he learned to speak up to Carol and say things like, "Alice and I are cooking dinner right now, and then we're going out. I'm sorry, it's not a good time to talk. When can I call you back tomorrow?" Bill needed to understand that Alice's negative focus on Carol had something to do with his failure to deal with her himself.

Taking Charge of His Daughter

Finding a voice as a dad was perhaps the most difficult challenge Bill faced, because Donna had become a high-maintenance, mouthy teenager. And for good reason. When Bill married Alice, Donna lost her special role as his caretaker, which she had so enjoyed while the two of them lived together weekdays after her parents divorced three years earlier. Now she found herself caught between two women who were parenting with a competitive edge, while her father made himself scarce. As Betty Carter points out, a teenage daughter is her mother's loyal torchbearer, with her stepmother as the natural target for her anger. Although Bill described Alice as "taking over," it was equally true that he had handed over the par-enting responsibilities to his new wife rather than stay in charge himself. Bill needed to understand that turning Donna over to the "woman of the house" was the surest way to put Alice in the role of wicked stepmother.

In contrast to Alice's strongly held beliefs about the right way and the wrong way to rear a teenage girl, Bill felt muddled, insecure, and uncertain about where he stood on various parenting questions. Before he married Alice, he had deferred to Carol—even after their divorce—when it came to anything concerning their daughter. He never even allowed himself to consider having a different opinion. Now, caught up in the tension between Alice and Carol and between Alice and Donna, he could no longer try to keep everyone happy by deferring to them.

In therapy, Bill worked to clarify his own beliefs about how best to rear his daughter. This allowed him to voice his opinions to Carol and also to take a position with Alice when appropriate. He said, for example, "Alice, I know how important neatness is to you, and I plan to work with Donna about her messing up the kitchen. I won't let her do that anymore. Let me work on getting her to be neater. But she's never made her bed, and she's always had a messy room, and I've always handled that just by closing the door. I don't think this battle is worth fighting, so let's let this one go."

Bill learned to discipline Donna, and to accept the hands-on job of daily parenting. He couldn't ask Alice to back off if he wasn't ready to move in. So Bill made sure that if his daughter needed transportation or anything else, he was prepared to arrange it rather than automatically expecting Alice to do it. Bill also let his daughter know that she didn't have to *like* Alice, but she did have to treat her with respect—a rule he intended to enforce.

"You Need to Back Off!"

Bill also needed to ask Alice to stop overloading the emotional circuits with her negativity about Donna and Carol. ("I just can't stand the way Donna dresses! Why does Carol let her walk around in those short skirts and tight sweaters?"). But Bill hesitated to tell Alice to make this change, because he was afraid that disagreeing with her would lead to a fight.

A turning point occurred one morning as Bill was heading into a telephone call with his ex-wife to discuss their daughter's clothing allowance. He was already tense, because he believed that Donna needed less money for clothes than Carol was insisting on, and he anticipated a struggle. Alice commented, with considerable intensity, that Carol was controlling and never failed to manipulate Bill and spoil Donna. Bill felt flooded with emotions. He turned to Alice and blurted out, "Lighten up, for Christ's sake! Okay, you don't like Carol. I don't always like her, either. But she's the mother of my kid, and you need to back off because I need to work with her, and I'm walking around with my stomach in knots all the time!"

Alice yelled back, "You're trying to put a muzzle on me! I'm part of this family, too, and I won't stand for it!" Bill was tempted to respond as usual by withdrawing and avoiding a fight. But to his own surprise, he stayed in the conversation. He responded, "No, Alice, I don't want to muzzle you. I want to hear you out about Carol or anything else. But I just can't deal with your constant negativity about her! When you keep criticizing her, it makes it *harder* for me to deal with her—not easier."

Because Bill spoke out during a tense moment, his "communication skills" left something to be desired. Surely nothing would be accomplished if Bill blamed Alice for blaming Carol and Donna. Blaming the blamer doesn't help. But Bill had vented his feelings, an angry exchange had ensued, and no one had dropped dead or filed for divorce. That was a start.

Setting More Limits and Boundaries

Over time, Bill continued to let Alice know that she should tone down her unhelpful negativity. He'd say things like, "Please don't criticize Carol in front of Donna or roll your eyes when Donna reports something her mom said. Donna needs to have the best relationship with Carol that they can work out. Plus, criticizing Carol just sets you up as a target for Donna's anger, which isn't fair to you."

It was new behavior for Bill to define his limits and boundaries so straightforwardly. With the help of therapy, he became increasingly able to define his position. He learned to say to Alice, "Look, the way you talk to me when you get upset immobilizes me. It's not productive. I don't mean to blame you, but I can't participate in the conversation when I feel paralyzed. So let's work out a different way to talk." He persisted in telling Alice what he needed and how he experienced her input. ("I need you to stop criticizing or correcting me in public. I also want you to tell me your criticism in one short paragraph. I feel flooded when you go on for so long"). I told him that he deserved a medal of honor for his persistent efforts to change how he interacted with others when his lifelong pattern had been to accommodate.

Bill continued to negotiate with Alice about parenting concerns. He'd say, for example, "Alice, I feel so lucky to be married to you. I'm really interested in your input. But it's very hard for me when you take charge and just tell me how things should be done. I know you have great ideas about parenting, and I want to hear them. But it's also true that there are some things we see differently. I'm Donna's father, and I need to deal with her in a way that makes sense to me, even if I make mistakes."

I also supported Alice and helped her to understand that while everyone assumes that women will take care of children—including other people's children—the attempts of stepmothers to be in charge usually backfire. This is especially true when there is an adolescent daughter on the scene. In any case, nobody can walk into a family with a separate history of its own and become an instant mother. Alice was able to hang out on the periphery for a while, to let go of her plans to be one big happy family, and to back off from the idea of becoming a parent to Donna. If she can develop a parentlike relationship over time (which is far more likely with younger stepchildren), that's a wonderful gift—not a given. Alice's efforts to find her voice in her new stepfamily will fail if she tries to accomplish the impossible.

Strengthening Other Connections

Last but not least, I encouraged Bill to strengthen his connections outside his marriage and to open up conversations with friends and family members. He needed support. Also, his marriage would become a pressure cooker if it was his only focus. Finally, if Bill could discover his voice with members of his first family, he would feel more grounded himself and find it easier to speak with clarity and confidence with Alice, Donna, and Carol.

For example, Bill's relationship with his widowed mother was cordial but superficial, so here too he faced the challenge of warming things up and speaking up. True to societal gender roles, Bill kept working when she visited, while Alice entertained her. I encouraged Bill to take a few days off and arrange for some one-to-one time with his mother. He could hardly remember the last time just the two of them had gone out together. Nor had Bill recognized that it was *his* job to entertain his mother, and that taking time off from work during his mother's visit might mean a lot to her.

When it comes to their family of origin, men often tell me that everything is fine. *Translation:* The relationship appears calm because it's distant. I suggested that Bill *talk* to his mother. He thought he was talking to her, but he actually shared very little of himself. He also needed to be more creative in asking his mother questions, since "How are you doing, Mom?" predictably elicited the nonresponse "Pretty good."

Bill rose to the occasion. He took time off work when his mother visited and began asking her specific questions about her past, her current situation, and her concerns for her future. To his surprise, she welcomed the opportunity to answer personal questions. Bill asked about his dad, who had died when Bill was seventeen, and he expressed interest in learning more about the family. He treated his mother as if she had something of value to offer him, and discovered that she did.

I strongly encouraged Bill to initiate conversations with his mother about his divorce from Carol. Bill hadn't revealed anything to her about the problems in his first marriage, including the part both he and Carol had played in creating the distance between them that led up to her moving out. His mother had been terribly upset by the divorce but had never felt free to question him or express her concerns. Now Bill invited her to do so, by asking her how she had responded to the news of his divorce and remarriage, and if she had any current concerns. He told her how important she was to his daughter and how grateful he was that Donna had her as a grandmother.

Before these conversations his mother had reacted with a negative edge toward Alice. Perhaps her negativity had been driven by her feeling neglected by Bill, who had initially circled the wagons around his new stepfamily and then left his wife to deal with his mother. That's another way women may play out the tension caused when men hang out on the edges of a relationship instead of getting involved.

No Easy Solutions

I don't mean to convey that Bill simply shored up his assertiveness and parenting skills, then worked on family-of-origin issues so that everyone lived happily ever after. I saw every family member (including Alice, Donna, and Carol) together and separately over a period of more than a year before their family situation settled down and they achieved a comfortable integration—which was ultimately aided by Donna's departure for college.

As a senior partner in a large law firm, Bill had no problem giving orders at work. But for a long time, he actually felt sick to his stomach whenever he contemplated expressing his opinion or taking a stand with Alice, even though he loved her dearly. He could have followed the line of least resistance and deferred to her indefinitely. Or he could have moved to another city and left his daughter with Carol as a way to keep his new marriage calm. Fathers often lose touch with their kids in this way after divorce or remarriage, not because of a lack of love but because they feel helpless about how to deal effectively with the other significant adults in their child's life. Instead, Bill took the hard road and learned to speak up with all the key people in his family.

Divorce and remarriage are extremely challenging life-cycle stages. They have built-in potential for competition, jealousy, loyalty conflicts, and the creation of "outsiders" and enemies both within and between households. Bill's story illustrates how incredibly hard it is for men, in their roles as husband and father, to define a solid and responsible position in the invisible web of family triangles. Even in less complex family situations, men may take the path of least resistance, choosing silence over speech because "it's not worth the fight." But men who disappear into silence with the women in their lives pay an extremely high price.

And so do the women in their lives. Alice was not a "wicked stepmother," and Donna was not "an impossible teenager." Alice and Carol didn't become enemies simply through their own immaturity. Every family member was caught in a web of interlocking relationships, where each person's behavior sent a ripple effect throughout the whole system. That's how families work. When men lay low and refuse to step into the ring, it's the women in their lives who often end up slugging it out.

CHAPTER 12

Criticism Is Hard to Take

Criticism, like life itself, is not always fair. Sometimes it's dreadfully unfair or just plain wearing. Let's look at the challenge we face when we feel wrongly attacked or unfairly picked on. How do we find our voice when *we* are the one sitting in the hot seat? How do we clarify our own position with dignity and firmness, without getting defensive or attacking the other person—and without backing down? How do we stay connected when we feel like striking back? What's the point of doing so?

MOTHERS AND DAUGHTERS: A CRUCIAL CONNECTION

When Katherine called me for help, she was still reeling from her daughter's accusations. Dee, who was twenty-six years old, had recently visited Katherine over the holidays and confronted her on Christmas Eve with a list of old grievances. Dee's timing wasn't great, and her message was devastating.

Dee accused Katherine of having been a selfish and self-absorbed mother who was responsible for Dee's problems with men and her bleak view of her own future. Dee also blamed her mother for her father's drinking after their divorce, when Dee was nine. Dee mentioned that she had uncovered these insights in therapy, which Katherine just happened to be paying for.

When I first saw Katherine, she told me that there had been zero communication between them since this angry confrontation. She herself had no plans to initiate contact because she was waiting for Dee to apologize for her "vicious attack."

A Word about Mother-Blaming

If you identify with Katherine, it may help you to know that many daughters blast their mothers somewhere along the way. Mothers often bear the brunt of anger for two, because daughters expect far too much from their mothers and far too little from their fathers. Also, mothers are less likely than fathers to disappear in the fray, and are therefore a "safe" target of attack.

Mother-blaming is still rampant in our culture, and this bad habit creeps into the deepest interior of family life. Mothers are held responsible not only for their own behavior (which is fair enough) but also for their children's behavior, which they can influence but not control. Mothers are blamed (and blame themselves) for every family problem. Many mothers are on the defensive even before they're attacked. Katherine was no exception.

But here's the real point: When it comes to family, some of the other person's complaints will be true, since we can't possibly get it right all—or even most—of the time. Only after we can hear our daughters' criticisms and anger, and are open to apologizing for the inevitable hurts and mistakes that every parent makes, can we expect to be truly heard by them. We need to be able to listen before we can get our own message across. That's actually good advice for every relationship.

A commitment to listening first doesn't mean that we sit in mute frustration. Obviously, we need to have limits. For example, Katherine might have said, "Dee, what you're telling me is really important. But it's Christmas Eve, and I'm finding it hard to continue the conversation just now. In order to give this my full and undivided attention, I'd like to wait until the day after Christmas."

Nor should we tolerate rudeness. We might sit still for the initial blast, but from what Katherine described, Dee's behavior was way out of line. So Katherine might have said, "Dee, I love you, but I can't listen to you when you yell or call me names." When we continue to tolerate rudeness in any relationship, we erode the self-regard of both parties. If we do call a halt to a conversation, we should approach the other person when we're calm and open up the conversation again.

How do we ensure that we will think clearly at the moment we are being criticized and blamed? We can't. What happened between Katherine and Dee was an escalating fight where nobody listened to anybody—and that's normal. What matters is what happens from this point forward, and whether Katherine responds to her daughter (whether through her words or silence) in a way that deescalates tension or intensifies it. Obviously, both mother and daughter have a lot at stake here.

In Praise of Thinking

Katherine was still shaking with rage when I met her. Her retreat from her daughter made sense. She felt overwhelmed, and needed to protect herself. Trying to continue the conversation would have been counterproductive until Katherine could handle her intensity. No one thinks clearly in the midst of a tornado.

To this end, I asked Katherine questions that would help put the conflict with her daughter into a larger context. Questions help us to think rather than just react. How was Katherine's relationship with her daughter similar to, and yet different from, her relationship with her own mother? What did Katherine know about her mother's relationship with *her* mother? Has there been a pattern of fighting or distancing between parents and children in previous generations?

What else was happening in her daughter's life or the life of the extended family that might have pushed Dee's anxiety higher around the time of her visit? How had Dee reacted at the time of the divorce, and had her thinking changed over time? How much talking had Dee done with each of her parents to help gain some perspective on the divorce? What conversations had Katherine initiated with her daughter about the painful events of the past?

A snapshot of family interactions across generations showed that mothers and daughters had not fared especially well. Katherine's mother and grandmother, for example, did not speak to each other for almost eight years after a fight that occurred in the aftermath of Katherine's grandfather's death. Katherine's relationship with her own mother was calm but chronically distant.

Having an authentic voice requires us to think about how we want to navigate our relationships so that we don't automatically repeat family patterns on the one hand, or mindlessly rebel from them on the other. Was distance and cutoff a pattern Katherine wanted to follow? Did she want to pass this legacy on to her daughter? What were Katherine's own beliefs and values about what kind of mother (and sister, daughter, granddaughter, aunt, and cousin) she wanted to be? Did she have an image in her mind of what she wanted her relationship with Dee to look like, say, five years in the future?

Getting Back in Contact

Several weeks after starting therapy, Katherine began the session by saying, "I'm afraid I might lose my daughter. What can I do?" As she explored her options, she realized that her impulse to pick up the phone might easily lead to more conflict. Instead, she decided to write Dee a note.

Katherine was an attorney with a fine eye for detail, so brevity didn't come naturally to her. Her first impulse was to build her case—a step in the wrong direction if her goal was to lower rather than increase intensity. Ultimately, Katherine decided to buy a greeting card in which she put a short message to test the waters.

> Dear Dee,
> How are you doing? It's been difficult for me to get back in touch since your visit. I know it was painful for both of us. I've been trying to consider the things you said as objectively as possible. It's hard, because I get defensive when I feel criticized. But I'll keep trying. I'm thinking about you.
> Love, Mom

Dee didn't respond to Katherine's card, or more accurately, she responded with silence. The challenge for Katherine was to view Dee's silence as *information* about her level of anxiety and intensity and not to take it personally. She also needed to remind herself that substantive change in family relationships is a slow process, not a one-shot deal.

Hanging In

About ten days after mailing the first card to Dee, Katherine wrote to her again. I had suggested that she pick up the phone and call, but she didn't feel ready. She was wise to consider how much anxiety she could manage. Her letter went something like this:

Dear Dee,

I'm sitting here on the red couch wondering how you're doing. I've continued to think about your last visit. I'm sorry that things got so intense between us. I assume from your silence that you are needing more space for yourself at this time.

I appreciate the courage it must have taken to share your feelings so directly with me. I want to have the kind of relationship in which we can talk openly.

Since your visit, I've been thinking more about my relationship with my own mother. I was never able to tell her when I was angry with her and I never stood up to her. This made things go smoothly between us, but we had a pretty superficial relationship. Maybe that's why I felt so unprepared to handle conflict between you and me.

My mother once told me that she and her mother were always fighting. From the stories your grandma tells, it was one big war and they stopped speaking to each for almost eight years after my grandfather died. So she and I reacted to that bit of history by doing the opposite and never letting a difference arise between us.

As I think about these patterns between mothers and daughters in our family, I realize how much I hope that you and I can have a different kind of relationship. I've also been thinking about the people in my family who aren't speaking to each other. I can't imagine anything more painful than that happening between you and me. So let's try again when you're ready, and I'll do my best to listen.

Love, Mom

This thoughtful letter illustrates the principles of Family Systems Theory 101. Katherine focused only on herself. She didn't request or demand a particular response from her daughter. She broadened the frame around the issue of mothers and daughters. She stood for connection without getting preachy. She didn't overload the circuits by going on too long. She didn't push for contact before Dee was ready.

Dee did respond to Katherine's letter, albeit briefly. She wrote, "Thanks for your letter. I'm swamped with grading papers. Will write when I can." These three sentences were an important sign that Dee was not entrenched in a position of cutoff. Not long afterward, mother and daughter were back on speaking terms.

When Katherine started therapy, she was resolved not to connect until Dee apologized. An apology for Dee's rudeness would have been nice, but it wasn't going to happen. Besides, Dee's rudeness was less important than the fact that she had finally gathered her courage to talk to her mother about the divorce and its aftermath. If Katherine had stayed focused on getting her daughter to act first, nothing would have changed at all. In fact, if you want a recipe for failure in any important relationship, just dig in your heels and refuse to change yourself until *after* the other person shapes up.

Listening Differently

Katherine would have preferred to avoid her next step. At a calm time, she took the initiative to revisit the hot issues, with an eye toward better understanding Dee's anger. Katherine asked, "Dee, can you tell me the ways the divorce affected you?" "What was the hardest part for you back then? "What about now?" "At Christmas, you said I ignored you around the time of the divorce. Can you tell me what you remember about that?"

We naturally become defensive when a family member begins to criticize us. We listen to refute or correct what is unfair or wrong in their comments. Sometimes we need to decide in advance that we will try to listen differently—that *all* we will do is listen and ask questions that will allow us to better understand where the other person is coming from. We can save our defense for a future conversation.

Not everyone can do this. It's difficult to listen to someone's pain when that someone is accusing us of causing it. We automatically listen for the inaccuracies, exaggerations, and distortions. To listen with an open heart in order to understand the other person requires intention, commitment, and practice. It is a spiritual exercise, in the truest sense of the word.

Find What You Can Agree With

Katherine's next step is to apologize for the part of Dee's confrontation she can understand or agree with. If she can wrap her brain around only 2 percent of what Dee was saying, she can at least validate that 2 percent.

Dee accused her mother of "selfish neglect," especially during the years before and after the divorce. Dee was clueless about what her mother had actually been up against during that time of crisis. Katherine had been quite depressed after the divorce and was physically exhausted by her efforts to stay afloat financially. She felt she had done the best she could, yet she also knew that her attention to Dee had suffered. In considering Dee's feelings, Katherine was able to empathize with Dee's experience of neglect, and she mustered the courage to tell Dee how sorry she was that she hadn't been more available to her at this painful time.

Defining Our Differences

What about the things Dee said that Katherine believed were wrong, unfair, or totally off the mark? Katherine could just let some of them go. Addressing every injustice and inaccuracy is not necessary. But a couple of her daughter's accusations stayed with Katherine, and she had to speak to them. She took the first step to define her differences in a letter that was otherwise chatty, and included the following three paragraphs.

Dee, I've been thinking more about our last conversation. I have a much better appreciation of how painful the divorce was for you and how much it affected you. I was in so much pain myself that I didn't pay enough attention to you. Like I said, I'm really sorry. If I could go back in time, I'd try to do a better job of being there for you. I would not have stayed with your dad as you wished. But I could have handled things better.

Here's where we see things differently. I don't believe that I'm responsible for your poor choices of men and your pessimism about making a good marriage for yourself. I don't doubt that I've done things in the past that contributed to your problems. But I don't take responsibility for the decisions you make as an adult. I have more confidence in you and more hope for your future relationships than you may feel right now. Also, I don't take responsibility for your dad's drinking after the divorce. And as you know, when people divorce, they usually have quite different perspectives on what happened. But whatever happened between your dad and me, I can take responsibility only for my behavior, not for his. He was an adult and I'm only sorry he didn't get the help he needed.

> I know we may see these points differently, and I hope
> we can be open about the places where we may disagree. I
> can't wait to see you over spring break.
> Love,
> Mom

What started as a crisis—and what might have resulted in another eight-year cutoff—became an opportunity for both Dee and Katherine to learn more about themselves and each other. And each woman emerged with a stronger and more confident voice.

BLOCK THOSE INSULTS!

One woman who wrote to me at *New Woman* faced a dilemma of how to best deal with her husband's constant criticisms. Her dilemma is a reminder that being open to criticism isn't the same as tolerating insults. She wrote:

> I would like to know what you would do if you had a hus-
> band like mine who kept telling you that he thinks you're
> fat and that it bothers the hell out of him? I've gained 20
> pounds since I turned 40 last year, and I'm tired of hearing
> his critical jabs.

There are two difficult voice challenges here. This woman needs to let her husband know that she won't listen to rudeness, that it isn't acceptable. Yet she also needs to keep the lines of communication open about her weight or anything else and express her willingness to continue the dialogue.

If I were in this woman's shoes, I'd begin by saying something like, "It's hard for me to talk to you about my weight when I feel like you're putting me down or insulting me. But I'm interested in hearing your thoughts about it." I'd ask him questions, so I could learn more about his feelings and associations. What specifically bothered him about my size? Was he embarrassed by me? Did it affect how he felt about me in bed? Was he worried about my health? When he was growing up, who were the fat people in his family? How did he and others relate to them? I'd express genuine interest in hearing his views, while acknowledging that this might be a sensitive issue for both of us.

While I wouldn't tolerate insults, I'd try not to punish him for his honesty, because I wouldn't want him to conceal his real feelings and genuine distress. I would try to listen without defensiveness. I'd also be honest myself. I might say, "You know, my weight bothers me, too, and I'm having a hard time doing anything about it." Or, "This is how I understand the timing of my weight gain. How do you see it?" Or, "This is the size I am, I'm comfortable this way."

But if my husband kept criticizing my weight or telling me I was fat, I'd tell him to back off. I'd probably accomplish this with some combination of lightness and humor, along with the more serious explanation that his comments hurt my feelings and weren't helpful. If losing weight was my goal, I'd let him know him specifically what he could do to support me in this effort, and what kinds of comments made it harder for me.

If he continued his critical jabs, I'd keep taking the conversation to the next round. I'd pick a time when we were close and I wasn't feeling the slightest bit angry. Then I might say, "I've been puzzling over something. I've told you several times that it hurts my feelings when you take jabs at my weight. Yet you still do it. Is the problem that you don't *believe* me when I tell you this is painful and not help-ful? Or is it that you do believe me—and you do it anyway? Help me understand this." I'd let him know that while he has a right to his feelings, he shouldn't express them at the expense of mine. I'd be available to talk with him about my weight, if he approached me empathically and respectfully. But if his critical comments felt demeaning or unhelpful, I would draw the line. I simply wouldn't let it go on. Certain kinds of talking don't even deserve to be called conversation, and it's better to firmly—or affectionately or playfully—refuse to go there.

Ten Do's and Don'ts

A woman's magazine recently interviewed me on the subject of coping with criticism. I offered the following tips to remember when you're on the receiving end:

1. Listen attentively to the person who is criticizing you without planning your reply.

2. Ask questions about whatever you don't understand.

3. Avoid getting defensive. Don't listen in order to argue or refute. Instead, listen for the piece of criticism you can agree with, even if it's embedded in exaggerations and inaccuracies.

4. Apologize for that piece first.

5. Never criticize a person who is criticizing you. There may be a time to bring up your own grievances, but that time is not when the other person has taken the initiative to voice her own complaints.

6. Stay calm. Underreact and take a low-key approach when dealing with the other person. Anxiety and intensity are the driving forces behind dysfunctional patterns.

7. State your differences ("Here's the piece I don't agree with . . .") only after you can do so without criticizing, blaming, or putting down the criticizer.

8. Stop a nonproductive conversation that is occurring at your own expense. It's fine to say, "I need a little time

to think about what you're saying. Let's set up another time to discuss it." Or, "I feel diminished when you talk to me this way. It hurts my feelings." Or, "I need you to bring up just one criticism at a time. When you start bringing up the past or list one thing after another, I shut down and can't listen."

9. Speak to the really important issues—and let the rest go.

10. When you're in the grip of strong emotions, remember this reversal of an old maxim: "Don't just do something. Stand there!"

No one likes being the target of criticism, but a lot can be learned from the challenge. With practice, we can enhance our capacity to listen differently, to ask questions, to get our intensity down, to move toward rather than away from the other person. We can apologize for the part we can agree with and speak to the differences. As the story of Katherine and Dee illustrates, a genuine apology can be deeply healing and can repair a disconnection.

It helps to keep in mind that other people usually don't criticize us with the intention of doing harm. Rather, people criticize us for the same reasons we criticize them. They want to be helpful and contribute to our betterment. Or we have a trait, quality, or behavior that bothers them and so affects our relationship, and they really do need to talk about it. They may think the relationship can't move forward if we don't consider our behavior and apologize.

Sometimes the other person's criticism has everything to do with them and little to do with us. A person may be anxious and having a bad day. In such circumstances, we may want to stay light and simply step aside from the negative judgment rather than make an issue of it. A more difficult situation arises when others chronically focus on us in a negative, controlling, or judgmental way, like the husband who took jabs at his wife's weight. We've already heard what they have to say, so the challenge isn't to listen more. Rather, we need to say, "Enough!" without driving an important subject underground. We need to tell a partner to tone it down and find a different way to talk to us, like Bill did (chapter 11) with his wife Alice.

But sometimes we ourselves have a grievance, and we want the other person to hear us out. We may deeply crave an apology from an important person in our own life. Let's look at the matter of apologies, and the folks who can or can't voice them.

CHAPTER 13

An Apology?
Don't Hold Your Breath

I once witnessed an extremely moving moment between a mother and a daughter. I was Letty's therapist when she invited her twenty-four-year-old daughter, Kim, to join us for a session. Kim had been avoiding her. Something was obviously wrong, but Kim refused to talk about it.

When Kim was twelve, her father had entered her bedroom at night and molested her over a period of several weeks. Kim's mother hadn't known what was happening, but responded appropriately by getting the whole family into treatment after the facts came out. Kim's dad had died recently of a heart attack, and his death stirred everything up again, including her enormous rage.

When Letty invited Kim to join us for a therapy session, Kim refused. But a few months later, she agreed to come just once. When I asked Kim how she had been doing since her father's death, she launched into a terrible attack on Letty. Although I'm trained to be a calm presence in an intense emotional field, my own anxiety rose in response to the raw rage that Kim directed not at her deceased father, but instead toward her mother. On a purely emotional level, Kim blamed Letty for her father's behavior. She located the "betrayal" in the family between mother and daughter, which is not unusual for daughters to do.

I was about to intervene when Letty rose from her chair and pulled it closer to Kim's. I thought she was going to yell back at her daughter: "How dare you say this to me! How can you blame me for what your father did? How could I have known?"

Instead, Letty turned to her daughter in the most fully present way and said: "I'm so sorry, Kim. I'm so sorry I didn't know. I'm so sorry I didn't protect you. I'm so sorry that this terrible thing happened in our family. I'm so sorry that you didn't feel safe enough to tell me the truth." Then she started to cry. Kim put her arms around her mother, and they cried together.

I don't know how Letty was able to be there for her daughter in such a remarkably open, nondefensive way. Letty didn't say she was sorry because she believed it was her fault or she had been a bad mother. She had been clear about this fact with Kim during the earlier family therapy that focused on the sexual abuse. But now, in the midst of being totally blasted, she moved into a place of pure listening and offered her love. Her apology for being part of this painful history was heartfelt and deeply healing for her and her daughter.

Letty's apology was especially healing because it didn't include any postscripts. She didn't say, "I'm sorry, but you need to keep in mind I didn't know it was happening." Or, "I'm sorry, but your dad was a weak man, and I don't think he could help himself." Or, "I'm sorry, but I wish we could put this behind us and move on." She didn't even say, "I'm sorry and I hope you'll forgive me." Of course

Letty hopes Kim will forgive her. But a pure apology does not ask the other person to *do* anything—not even to forgive.

"I REFUSE TO SAY 'I'M SORRY' "

Letty apologized for actions beyond her control. In contrast, many people can't apologize even when—or especially when—they are culpable. They may desperately need to keep guilty feelings at bay, or to hold to a stance of innocence and goodness, to convince others and to reassure themselves. They may feel that admitting error and wrongdoing will open the floodgates and leave them vulnerable to bearing the pain and hearing the accusations of others till the end of time.

For countless reasons, some people won't ever apologize. Children who are badly shamed, blamed, and criticized while growing up may have a difficult time apologizing as adults. Or, if the adults in our life didn't apologize when we needed them to, we can find it hard to forge a new pattern for ourselves.

Alternatively, apologies are too big a deal in some families. One man I saw in therapy was seriously allergic to apologizing to anybody. "My parents were always in my face to get me to apologize to my brother, and they always assumed everything was my fault." His folks would say, "Now you apologize to Scott right now!" Then, "That wasn't a *real* apology. Now say it like you mean it!" He found the whole process so humiliating that his solution as an adult was to never say he was sorry. If other persons felt he owed them an apology, he would withdraw into silence or dig in his heels.

Some people are so hard on themselves for the mistakes they make that they don't have the emotional room to apologize to others. "I feel badly enough about what I did, and I want to move on," Deborah told me in therapy. She had missed her younger sister's wedding because it conflicted with a professional conference where she was presenting a paper. The conference had been scheduled before her sister announced the wedding date, and Deborah was initially angry at her sister for expecting her to be there. But the day of the wedding, Deborah felt awful about the choice she had made, and she wished she were with her family at such an important time. Indeed, she felt so bad that she didn't want to "open the whole thing up" or "make a bigger deal of it" later by apologizing. The thought of revisiting the issue with her sister only made Deborah feel more vulnerable.

Several years later, in a moment of sudden affection for her sister, Deborah spontaneously sent her an e-mail that said: "I've never told you how bad I felt about missing your wedding. The day I was giving my paper at that conference, I kept thinking to myself, WHAT AM I DOING HERE? I have no explanation or excuse for making such a stupid decision." Her sister wrote back, "Yes, Deb, you were a real asshole." They never discussed it further, nor did they need to. "It feels lighter between us," Deborah told me. "It's like some bit of trust has been restored that I didn't even know was missing."

People who apologize easily feel empowered by saying, "I'm wrong, I'm sorry, I apologize." An apology leaves them feeling better. People who apologize with difficulty have a different experience and belief system. An apology is associated with feeling worse, with weakening the fabric of the self and the relationships they depend on, with losing power or control. In Deborah's case, the act of sending the e-mail allowed her to correct an old belief system ("revisiting an issue makes things worse") that may have protected her in the past but no longer served her well.

"I'LL ONLY APOLOGIZE FOR *EXACTLY* MY SHARE!"

We all have idiosyncracies about apologizing. One of the quirks I share with my husband is that I like to apologize for *exactly* my share of the problem—as *I* calculate it, of course—and I expect him to apologize for his share, also as *I* calculate it. We're not always of one mind regarding who is responsible for what, which can lead to the theater of the absurd.

"I apologize for my part of this fight," I announce to Steve. I have appeared at his office door to make reparations. We just had a stupid argument, and I can tell Steve won't talk to me until I apologize.

"And what do you think your part is?" he asks testily.

"Forty percent," I say. True, I approached Steve in a critical way, but he was reactive and blew the whole thing out of proportion. I've concluded that his reactivity is definitely worse than my initial rudeness.

"Well," Steve retorts, "that's not good enough." Round and round we go. We sound like idiots, even to ourselves.

Fortunately, such a scenario only happens on our more immature days. Usually Steve or I will cut right through the nonproductive "whodunit" or "who started it" mentality. One of us will say, "Let's stop this. I apologize for my part in what's going on." The other accepts the apology, and that's that. If one of us is in a light mood, the other can't push the conversation downhill.

"WHY SHOULD I APOLOGIZE FOR *THIS,* WHEN SHE DID *THAT?*"

Like Steve and me, many folks have a harder time apologizing if they feel "overaccused" or pushed to assume more than their fair share of the blame. As one man put it, "When my wife criticizes me, I don't want to apologize because I feel like I'm sticking my head on the chopping block. It doesn't feel equal. If I apologize, I'm agreeing with her that I'm the whole problem. And that's not true." If we experience offering an apology as a blanket statement of our culpability or one-downness, we will feel a loss of power and esteem and find it difficult to express how sorry we are for our own contribution to the problem.

A Painful Impasse

Consider Jayne and Denise. They now live on opposite coasts but share a long history of friendship. When Jayne adopted a baby from China, she neglected to inform Denise right away that the adoption had gone through. Denise heard the news from a mutual friend and felt upset to be treated like an outsider. She sent Jayne a scathing e-mail, demanding an apology for being left out of the loop on such important information.

Jayne is the sort of person who has trouble admitting error and apologizing in any situation. So Denise's attack, and her insistence that Jayne owed her an apology, made the situation worse. Jayne didn't answer the e-mail, because she experienced Denise's criticism as finger-wagging that exaggerated a simple oversight and asked her to grovel.

Jayne told me she felt controlled by Denise's anger and her demand for an apology. "I don't owe Denise anything!" she insisted. "If she had stayed in better touch with me, of course I would have mentioned the adoption. But I've done most of the reaching out in this relationship, and I'm sick of it. I've visited her twice in Seattle, but she's never come to Philadelphia. Also, I simply forgot to send her an adoption notice. I didn't plan to leave her out." Jayne believed Denise owed *her* the apology and needed to admit her part in creating the situation.

Jayne also thought Denise was making a huge deal out of nothing. On a 1-to-10 scale of bad behavior, Jayne believed her oversight barely scored. Yet Denise was treating it like a deliberate affront. This made Jayne even more defensive, as if apologizing would force her to take more blame than she deserved.

Denise, for her part, had a deep need to hear an apology whenever she felt wronged. She had experienced many painful events growing up, and her parents had never acknowledged her experience, affirmed her reality, or accepted responsibility for their behavior. As an adult, Denise was guided by a strong sense of justice and a great need to have her reality validated. When she felt mistreated, she wanted to be told: "Yes, this did happen. I was wrong. I'm sorry I hurt you."

With their friendship at an impasse, Jayne sought my perspective. I encouraged her to take the high road and deescalate the situation by being the first to apologize. When she thought about it, she realized she should have sent Denise an adoption announcement—whether or not Denise was holding up her end of their friendship. So she sent Denise an e-mail that said, "Denise, I really am sorry I didn't send you the announcement about my baby. Let's talk! Love, Jayne." Denise felt relieved and much more at peace. The two women ended up having a good talk about their relationship, in which both were able to voice their dissatisfaction. If Jayne had stayed defensive, a minor conflict might have turned into serious disconnection.

THE CHALLENGE OF DIFFERENCES

We all have deep-seated beliefs about giving and receiving apologies, even if we don't articulate them. Such beliefs are rooted in family and culture, and may be generations in the making. Some cultural groups place a high premium on apologies and forgiveness. Others don't. When we begin to expose our beliefs to the light of day, we can look squarely at how well they work for us, and we can revise them if necessary.

An e-mail from my nephew, Yarrow Dunham, reminded me about the ever-present dilemma of differences in relationships, in this case about apologies. Yarrow is living in Japan and has been dating a Korean woman, Kyong Nam, for several months. Yarrow wrote:

> Our most recent collision involves the apology. In the after-math of an agonizing conversation, full of sundry twists, turns and accusations, she told me that she has never, not once, apologized to her mother, father or a lover(!). For us, used to invoking the apology even for things completely outside our control, and forged by parents and teachers who demanded stand-up formal apologies, and "it better be sin-cere," the fact that she'd never apologized grated. Seems the Korean idea (yes, I dare generalize from one woman to an entire culture) is that, to the degree our relationship is intimate, we don't need to apologize. Of course the person will forgive us, and of course they can read the non-verbal signs of apology, and of course they know the non-apologizer recognizes their own wrong and will try not to do it again. Apology is a social activity that marks distance, a lack of intimacy, because if intimate you wouldn't need it. Well, it's good to know.

For Kyong Nam, love quite literally means never having to say you're sorry. If two people in a relationship have no investment in offering or receiving amends, it need not be a problem. But fre-quently an apology really matters to one person, and the other can't or won't offer it. After all, every relationship is a cross-cultural experience of sorts. The culprit might be the person who won't apologize, or the one who doggedly demands apologies, or it might even be the tug-of-war between them. How you see it is influenced by your belief system and with whom you happen to identify.

GOOD APOLOGIES AND BAD APOLOGIES

A psychiatrist friend tells me flatly, "I don't apologize and I don't accept apologies. When people apologize to me, they're trying to silence my anger. They're really saying, 'Look, I apologized, so be quiet already. Drop it.' "

My colleague views the apology as a manipulative tool to silence or placate the other person, and to grab the moral high ground. As I listen to her point of view, I'm struck by our stark difference of opinion. When someone offers me a genuine apology, I feel relieved. Whatever anger and resentment I still harbor melts away. I also feel so much better when I offer an apology I believe is due. It's important to me to know I can make mistakes or act badly, and then repair the disconnection. Without this possibility the inherently flawed experience of being a human being would feel impossibly tragic. Tendering an apology, beyond the social gesture, can restore a sense of well-being and integrity to those who sincerely feel they did something wrong.

But my friend and I do share some common ground. We both agree that women, in particular, often apologize to a fault. Indeed, many women feel guilty if we're anything less than an emotional service station to others. Or we may be just as quick to feel responsible for everything. My friend Meredith tells the story of pausing on a ski slope to admire the view, only to be knocked down by a careless skier who apparently didn't see her. "I'm s-o-r-r-y," she reflexively yelled after him from her prone position as he whizzed by.

Apologies That Don't Heal

Not every apology leaves both parties feeling better. An apology can be offered without sincerity, simply to get out of a predicament. Or the wording of the apology simply lets the person off the hook. You might not be satisfied to hear your partner say, "I'm sorry you were so upset by my comment at the party," as if the problem is that *you're* oversensitive. You may want to hear, "I'm sorry I criticized you for telling jokes at the party. I was out of line to act like it was my job to monitor your behavior."

Then there are apologies followed by rationalizations. We've all heard these (or offered them), and they are rarely satisfying. My friend Jennifer Berman drew a cartoon of the "guy with a million excuses." My personal favorites were, "Hey, I TRIED to call but all of a sudden my ARMS SHRANK and I couldn't reach the dial—honest!" and also, "I'm sorry . . . but you never ASKED me if I was married with kids." When we confront someone about something they did or didn't do, it can be so satisfying if they can just say, "I'm really sorry," and leave it at that. We ourselves can practice apologizing ("Mom, I'm sorry it's been so long since I called") without the "add-ons" implying we're not really responsible for our actions ("but my work has been so overwhelming I didn't have a free second").

Some folks apologize with a grand flourish but then go on to continue the very behavior they are apologizing for—whether it's drinking or coming home late from work after they've agreed not to. An endless, meaningless string of apologies signals a failure to change one's own behavior. What matters is not whether the person was authentic or "really meant it" in those passionate expressions of remorse. All that counts is whether that person follows through so there is no repeat performance.

Some apologies serve mainly to silence the other person, as my colleague noted. A man who started therapy told his wife six months after she discovered his affair, "I *said* I'm sorry, so why are you bringing it up *again*?"

Possibly this man needed to set limits if he felt that the affair was repeatedly being thrown in his face in an unhelpful way. In the future, he might have to say, "Look, it's not constructive for you to keep bringing up my affair every time you're upset or mad at me. I'm committed to talking to you about it—but not this way. Let's plan a time to talk." In contrast, it wouldn't be fair of him to demand, "Stop! I've apologized for the affair fifty times and I refuse to ever talk about it again." Nor would it be fair to expect that she would "just get past it," or always control her rage. No apology can substitute for a willingness to hang in there, and engage with the other person over time. Nor can saying, "I'm sorry," substitute for the hard work of healing a betrayal and doing whatever is necessary to rebuild trust.

A Burdensome Apology

Finally, an apology can disempower the other person if we don't have our own emotionality in check and we *overdo* it. One mother ran a stop sign and hit another car, with her sixteen-year-old daughter in the passenger seat next to her. The mother escaped with minor bruises, but her daughter sustained serious injuries that required two surgeries and a long period of rehabilitation. Of course, the mother was beside herself with guilt, grief, and remorse. She also dreaded her daughter's unspoken anger. Several times a day, this mother would tell her daughter how sorry she was and how she would never forgive herself. When the daughter expressed emotional or physical distress, the mother would say, "If only it was me, not you. I'd give anything to change places with you!" The daughter began to feel angry, crowded, and disempowered by her mother's constant focus on how painful it was to see her suffer. "Enough already!" she shouted at her mother one morning. "This is *my* suffering, and I'll deal with it. Go take *your* suffering someplace else." The mother, to her credit, got the message and entered therapy to cope with her guilt.

Perhaps you can think of apologies or expressions of remorse you would have preferred not to receive. Perhaps a person who has harmed you in the past decides she needs to apologize to facilitate her own recovery, without considering whether dredging up very old material might be more painful than helpful for you. Or perhaps you can recall an incident where you might have done a better job yourself on the giving end. Not all apologies clear the air or foster the growth of the relationship.

ADDRESSING THE UNSPEAKABLE

I vividly recall a brief exchange that occurred many years ago during a dinner party at my friend Maureen's house. The guests were part of the mental health community, except for a Japanese woman who was in Topeka because her psychiatrist husband had come to study at the Menninger Clinic. I hadn't met her before, but the rest of us were friends. I have no idea how it came up, but at one point Maureen turned to our guest and told her how deeply sorry she was about the bombings of Hiroshima and Nagasaki. She said, "I hope you know that many Americans think that dropping the bombs was a terrible thing."

Maureen's apology sounded so personal and so heartfelt that I was startled. I remember thinking to myself that Maureen hadn't even been born when the bombing occurred, and the rest of us had been in diapers. Yet her apology felt right. Her words were followed by a brief silence at the table. Then the Japanese guest said simply: "Thank you for saying this. My husband and I have been in this country for some time, and no one has said this to us before."

I felt moved by this brief exchange, just as I felt moved in my consulting room when I witnessed Letty saying to her daughter, Kim, "I'm so sorry that this happened in our family." The conversation led me to think about how some people who are not culpable can apologize in ways that are healing, but the people who *commit* great harm often cannot.

"It's Not My Fault": Minimizing, Rationalizing, and Denying

One might argue that a person who commits any terrible deed may lack the empathy and compassion necessary to feel remorse. But this is not always true. More to the point, the human capacity for self-deception is extraordinary, sometimes also adaptive and life-preserving when the facts are experienced as unmanageable.

Whether the context is personal or political, all of us can create layers of defensiveness and denial when we have done harm. One tells oneself, "It wasn't my fault," or "I couldn't help myself," or "It was necessary," or "It's not that big a deal." The more serious the harm, the deeper the levels of self-deception that come into play, and one tells oneself, "She really asked for it," or "I didn't do it," and even "It never happened."

The more unmanageable the guilt and shame, the more difficult for the wrongdoer to admit to harmful behaviors, empathize with the harmed party, and feel remorse. Instead, we hear rationalizations, denial, and self-serving explanations that shift the blame to the harmed party.

The most urgent issues—those where we may feel most desperate to be heard and understood—pertain to violations of trust by people we love and rely on. The process of opening up conversation on such subjects is slow and painstaking. When the violations are serious, professional help is necessary. The wrongdoer needs treatment in order to have the opportunity to feel responsible and accountable, and to move forward from there. The harmed party needs support, perspective, a strategy, and a long-range plan so as not to be retraumatized by trying to open a painful conversation and speak her own truths.

THE WRONGDOER MAY NEVER ACCEPT RESPONSIBILITY

A man named Ron punched his wife, Sharon, in the face and stomach. The second time this happened, Sharon filed a police report.

It's easy to empathize with Sharon's need to hear Ron apologize, to hear him say that he is fully accountable for his actions, that he is deeply remorseful, and that he will do whatever is necessary to rebuild their relationship and to ensure he never hits her again. It's far more difficult to empathize with Ron's failure to "get it"—with his resistance to hearing Sharon insisting that he take responsibility for the harm he has done.

I invite you to put on your therapist's hat (we're all psychologists at heart) and try to understand Ron's experience. People who commit serious harm can't be reached through conversations that further shame or blame them—nor by conversations that let them off the hook, for that matter. The following discussion may help you understand why you need to reduce your expectations to zero if you decide to speak out to a person who has harmed you, and why anger does not get through. Even as we appreciate the wrongdoer's humanity, we need to understand that people can't be more honest with us than they are with themselves.

Treatment: Inviting the Wrongdoer to Accept Responsibility

Ron was referred to me for therapy and also to a group for batterers. During our third court-ordered therapy session, Ron told me that he couldn't stand being in "a group of batterers." I asked him if he could stand being in a group of men who had problems with their anger spilling over. Could he see any benefit to getting a few tools that would help him become stronger and more powerful than his anger—so that his anger wouldn't control him?

"Definitely," he said. He would like to join such a group.

"What's the difference," I asked, "between a group for batterers—and a group for men whose anger is spilling over?"

"All the difference in the world," he said.

Ron was resisting the notion that his crime defined him. You might argue that Ron *is* a batterer, and that any language that softens or obscures this fact leaves him less accountable for his actions. But Ron will be more likely to accept responsibility and feel remorse if he can view himself as more than a batterer. For people to look squarely at their harmful actions and to become genuinely accountable, they must have a platform of self-worth to stand on. Only from the vantage point of higher ground can people who commit harm gain perspective. Only from there can they apologize. Of course, even a heartfelt apology cannot right a terrible wrong, but it can acknowledge the consequences of a wrong and clear the space for further healing.

If Ron can feel *more* than his violent behavior, he may talk in treatment about where he learned to behave as he did. He may be able to recall incidents in his life where he believed, felt, or acted in ways he can identify as good and honorable, whether as a son, brother, husband, neighbor, or breadwinner. He may be able to examine what he has been taught about "being a man" and about how women should be treated. He may consider whether his aggressive actions truly reflect his core values and beliefs about the sort of man he wants to be. He may begin to formulate a picture of the relationship he hopes to have with Sharon several years from now. He may apologize to her in a deeply heartfelt way. He may agree to make reparations for his behavior, for example, by working on Saturday mornings and giving the money to a "safe house" for women. When one has harmed or betrayed another person, saying "I'm sorry" is not enough.

Most people who commit serious harm never get to the point where they can take all or even some of these actions. But it's not just an expression of Ron's pathological denial that he doesn't want to label himself "a batterer" or "an abuser." To refuse to take on an identity defined by one's worst deeds is a healthy act of resistance. If Ron's identity as a person is equated with his violent acts, he won't accept responsibility or access genuine feelings of sorrow and remorse, because to do so would threaten him with feelings of worthlessness.

We cannot survive when our identity is defined by or limited to our worst behavior. Every human must be able to view the self as complex and multidimensional. When this fact is obscured, people will wrap themselves in layers of denial in order to survive. How can we apologize for something we *are,* rather than something we did?

Beware of Psychological Rationalizations

Nor can the wrongdoer maintain honor and dignity when allowed to rely on excuses and psychological rationalizations. A *New Yorker* cartoon shows a woman seated in the witness box saying, "I know he cheated on me because of his childhood abuse, but I shot him because of mine." Psychological explanations are not helpful when they invite people to avoid being fully accountable for their actions and for the harmful consequences of their decisions.

Of course, we all need to consider how the past and the present affect us. But a difficult past or painful current circumstances don't *cause* violent or irresponsible behavior. Many people who have suffered a traumatic past and horrific present circumstances do not go on to harm others. If Ron is viewed as a man who does not have agency, choice, and will, he will also lose the opportunity to change and be truly accountable for his violent behavior.

Ron was fortunate to participate in a comprehensive treatment program that did not label him as bad or sick, that enhanced his self-respect while holding him accountable for his harmful behaviors. He was ultimately able to accept the invitation to feel responsible, to make reparations, and to apologize to Sharon in a way that had meaning. But without treatment, it would be unrealistic to predict that he would make amends to Sharon—no matter how passionate and articulate her voice. It's difficult enough for the most comprehensive treatment programs to get results.

Speak Only Because You Need To

Even if you speak with love and respect—recognizing the wrongdoer's full humanity beyond the bad deeds—he may still never hear you. So don't speak because you need an apology or validation. Rather, speak to focus on what you want to say about yourself, for yourself. Longing for a genuine apology or an affirming response is totally understandable, but unrealistic when you enter a conversation with someone who has betrayed you. The only reason to speak is because you need to speak.

If you decide to open up any painful and loaded issue, let go of any expectations of getting the response you want. We will always come from a more solid place if we speak to preserve our own well-being and integrity and refuse to be silenced by fear—*not* because we need a genuine apology from the other person or expect to have our reality validated. Neither may be forthcoming, now or ever.

So keep these points in mind: No individual will feel accountable and able to apologize—no matter how we communicate—if doing so threatens to define her in an unacceptable or intolerable way. The other person's willingness to own up to harmful deeds has nothing to do with how much she does or doesn't love you. Rather, the capacity to take responsibility and feel remorse is related to how much *self*-love and *self*-respect that person has available to draw on. We don't have the power to bestow these traits on anyone but ourselves. We can only keep the other person's full humanity in mind and never forget that every human being is better and more complex than the worst things he or she has done.

CHAPTER 14

Complaining and Negativity: When You Can't Listen Another Minute

"Harriet, you must be so tired of hearing me complain," a dear friend tells me on the phone. "No, no! Not at all!" I reassure her—and I really mean it. The word *complaining* has pejorative connotations to almost everyone but me. I like complaining to my friends and I like being complained *to* (although, admittedly, I don't enjoy being complained *about*). I rarely think to myself, "Oh, no, here she goes again. I've heard this a hundred times already." I know that problems are complex, that life is one thing after another, that some bad things can't be fixed in a day—or at all. When I love someone, I want to hear the good and the bad, and I expect the same in return.

Of course, the complaining of some folks may exceed the limits of even the most sympathetic ear. A story by Yiddish scholar Leo Rosten illustrates the meaning of the word *kvetch* ("to fuss and complain with audible sound effects").

In the old days, passengers in railroad sleeper cars occupied upper and lower berths. A Mr. Fortescue, tossing and turning in an upper, could not get to sleep because from the berth below came a woman's constant kvetching: "Oy, am I toisty. . . . Oy, am I toisty!"

On and on went the lament, until Mr. Fortesque got out, crawled down the ladder, padded the length of the car, filled two paper cups with water, brought them back, and handed them in through the curtains to the passenger in the lower berth.

"Madame, here. Water!"

"God bless you, gentleman; thank you."

Fortescue crawled up into his berth, and he was on the very edge of sleep when, from below, he heard, "Oy, vas I toisty . . ."

I confess to being a bit of a kvetch myself. I'm not known to endure even moderate hardship without making a fuss and calling on my best friends and family for sympathy and hand-holding. No one will say at my funeral, "She was so noble. She never complained." Similarly, I want my loved ones to feel free to come to me with both their minor complaints and their major crises, even if I can do nothing more than sit quietly and bear witness to their pain.

But we all have our limits on how available we are to listen, just as we all have our limits on how much we can do or give. We may feel weighed down by the other person's grousing, which can take up too much space in the relationship and feel formulaic—like a tape automatically going round and round—rather than truly felt. If that other person happens to be a family member, rather than a *kvetching* stranger on a train, our own mood may spiral downward in response to these chronic expressions of worry or negativity. When our capacity to listen has been exceeded, we need to find a way to end the conversation or move it in a different direction. The goal is to protect the self without acting at the expense of the other.

"On and On, Anon"

What is a complainer? We tend to apply this pejorative label to anyone who voices problems in a manner that elicits our irritation rather than our sympathy. He or she may listen poorly and lament endlessly. "I'm going to refer my mom to that Twelve-Step group for people who talk too much," a therapy client quips, "On and On, Anon." Another client tells me, "I know my father is grieving for his wife, but it's all I hear about, and he's totally centered on his misery. I try to be a good son, but after he repeats the same thing for the two-hundred-thousandth time, I want to tell him to get over it." If the other person rejects our best efforts to help, and takes no positive action on his or her own behalf, the challenge of compassionate listening is especially great. It may be difficult for the listener to get past the need to be helpful and to accept the reality that the complaining party is not able or willing to take steps to solve a problem or to move out of a negative space.

Compassionate Listening

No how-to tip captures the quality of pure attention that occurs when we listen best, when we are fully emotionally present without judgment or distraction, when we are fully open and receptive to what the other person is saying without having to change, fix, correct, or advise, when we are *there* with that person and nowhere else.

We all are capable of much deeper levels of listening than we may ever tap into. Once, a moment of deep listening came to me when I least expected it, taking me by surprise. I participated in a two-week workshop on transformation and spiritual growth, held in the Arizona desert and led by an extraordinary teacher, Carolyn Conger. As we sat in a quiet circle, one woman shared her profound sense of isolation and despair. As she spoke, I felt fully present with her and with the others in the group in a way I had never experienced before. Compassion, connection, detachment, and appreciation for the sacred came together in this pure moment of listening and unconditional love. We have only to experience something once to know of its absence in our lives—and its possibility.

We can all improve our capacity to listen, and it's well worth the effort. Listening well is at the heart of intimacy and connection. When we are able to listen to another person with attention and care, that person feels validated and enhanced. When we enhance the other person, we also enhance our own self. Surely human consciousness would take a big leap forward if our wish to hear and understand were as great as our wish to be heard and understood.

But in everyday life, we can't always open our hearts to the other person's lament and offer the gift of our attention. We also need to figure out what to do at those moments when we just can't listen anymore, when we're feeling stressed, tense, and preoccupied ourselves, or when we are just plain irritated at hearing more of the same. If we're feeling raw, it's an act of self-love to protect ourselves from hearing the same thing even one more time. And as we saw with Janet and her sister Belle (chapter 4), there's nothing compassionate about letting a person go on and on when our limit to listen has been exceeded, or to only listen and never share our own problems or pain.

Countless factors can make us want to stuff a sock in the other person's mouth or stick our fingers in our own ears. Adult daughters are especially sensitive to complaining mothers—and understandably so. The relationship between mother and daughter is never simple, and a daughter often has trouble sorting out where responsibility to her mother ends and responsibility to herself begins. Whether we believe that our mother gave us too much or not enough, it's painful to be confronted with her unhappiness and to feel that nothing we do makes any difference.

A HOT SPOT WITH MY MOTHER

Here's an example of my efforts to be both open and shielded with my mother, Rose. When my parents were both in their early eighties, they relocated to Topeka to be near me and my family. My father moved to a nursing home (now called a health care center), and my mother into an independent living apartment. Throughout their long marriage, Rose had watched over Archie, but now she was helpless to monitor even the simplest details of his care, such as whether a blanket had fallen off his bed during the night, or whether the window shades had been opened in his room in the morning to let in the sun. Money had always been a persistent source of worry for Rose, and what would prove to be nearly a decade of care for Archie in three different nursing homes was eating away at resources she had worked incredibly hard to accumulate.

My mother had multiple problems of her own, but it wasn't her way to speak about them directly. During her many years in Topeka, I can't recall her once picking up the phone to say, "I'm having a bad day. Can I come for dinner?" or, "Can you and Steve come visit? I'm bored." She had pushed her needs aside for so long that she barely recognized them; plus, she didn't believe in "burdening" her busy daughter.

At times of stress, my mother has always focused on my father—be it in a critical or worried way—so it was no surprise that she became riveted on Archie after their difficult move to Topeka. Despite her enormous love of life, Rose was alone in her apartment, without sufficient connections, activities, or purpose, and so she began to focus narrowly and intensely on my father. She became increasingly obsessed with possible billing errors, or with what was being done to Archie or not being done for him. She lost objectivity and balance, focusing only on the negative, and she lost sleep ruminating about the cost and quality of Archie's care. Steve and I were the only ones she turned to with her escalating anxiety and distress. So intense was this focus in my otherwise mellow mother that at times I'd hold the phone away from my ear and try to concentrate on my breathing.

I never had the hubris to imagine I could handle my mother's situation any better than she did—only differently. We all have our own style of managing unrelenting stress. The problem for me was that I wasn't shielded from the intensity and tenacity of my mother's focus on Archie. After about two minutes of Rose anxiously insisting that Archie couldn't possibly have used as many wipes, pads, and diapers as a particular bill indicated, my mother and I were like two nervous systems anxiously twitching together.

My automatic responses only made things worse. When I felt especially allergic to hearing about Archie, I'd temporarily distance from Rose by calling or visiting less frequently. This kicked my mother's anxiety higher, so her focus on my father became even more tenacious. Sometimes I'd try to "reason" with her and explain the position of the nursing home or billing system. Rose, in response, felt misunderstood and without an ally, so she would then redouble her efforts to make her point. Sometimes I'd don my therapist's hat and suggest that perhaps other problems were being obscured by making Archie her full-time job. At my worst, I'd snap at her. "I just can't listen to any more talk about Daddy today!" I'd say. Or, "So *what* if they charged him for an extra box of diapers! Is it really worth being miserable over?" In response to the obvious irritation in my voice, Rose felt criticized, and nothing positive was accomplished.

I knew from my work with families that intensity only breeds more intensity, and that reactivity breeds more of the same. But it was almost impossible for me to lighten up. In my mature moments, I didn't feel judgmental about how my mother was handling her situation. How does one handle such a situation? Nor did I want to turn away from hearing her pain. But her relentless focus on my father felt less like a real sharing of feelings and more like a primitive flow of anxiety going from her body into mine.

Getting Creative (with a Little Help from My Friends)

On my good days, feeling calm and centered myself, I could be patient and empathic with my mother, no matter what. I could listen with an open heart and say, "The situation with Daddy is so terrible and has gone on for so long. I don't know how you're surviving. Is there anything I can do to help?"

On my not-so-good days, I'd grab Steve or a clear-thinking therapist friend by the collar to help me get a grip on my intensity so I could limber up my brain and respond more creatively to Rose in defining the limits of what I could listen to. I don't always do my best thinking about my own family (compared to, say, someone else's family), and if I'm drowning in emotions, I may not think at all. That's when I get some coaching on how to approach an old conversation in a new way.

Speaking from a Loving Place

One day, for example, my mother's focus on Archie and his nursing home bill was over the top. I sat down on the couch next to her, gave her a big hug, and said, "Mommy, I love you dearly. Everyone should have a mother like you. I've had so much fun with you, and I've learned so much from you. I'm sorry that things have been difficult for so long. But recently I've felt as if I've lost my relationship with you. It feels like when we're together we talk about Archie 98 percent of the time, instead of talking about each other."

My mother's first response was understandably defensive. "Then I won't mention him again!" she said. "If you're not interested, I won't talk about it." My natural tendency was to get defensive in turn, but I stayed on track.

"I *do* want to know what's going on with Daddy," I continued warmly. "It's not that I don't want to be supportive. I know it's painful for you to cope with Daddy's condition. But I feel like the amount we talk about him is out of balance."

"What do you mean?" my mother asked.

"I just think Archie is getting way too much of his share of our attention," I explained. "I feel that talking so much about Daddy's illness and problems has taken my mother away—and I'm feeling sort of sad and lonely about it."

As the conversation progressed, I felt that my mother really began to hear me, and that she allowed herself to be affected by my words, because they were coming from a loving rather than a critical place. I then shifted the conversation to other subjects, eliciting my mother's stories from the past and asking her opinion on decisions I had to make. In this way, I moved Archie off center stage. Sometimes the other person may be at a loss about how to stay connected except through automatic complaints (or, in some cases, criticism and advice-giving). When we steer the conversation away from the habitual, we need to offer other avenues of connection.

There's Not Always an End Point in Sight

In real life, entrenched family patterns don't change after just one or two conversations. In fact, a deeply grooved pattern is unlikely to disappear entirely, but may instead reinstate itself at times of stress. My mother's tendency to move toward me with an intense focus on my father began in my teens, long before he became old and ill. So the challenge for me was ongoing. When Rose focused on Archie in an anxious way that surpassed my limit for the day, I had to be creative to manage my own intensity.

Humor helped. I learned to laugh and tease Rose a bit. "Mommy, I do believe you're getting a little obsessed here. If you mention Daddy one more time, I'm going to come right over and give you a poke!"

I also spoke to the differences between us, taking care not to criticize, blame, or try to change her. "You know, Mommy," I said to her more than a couple of times, "I think we're in very different places when it comes to Daddy's bill from the nursing home."

"How so?" she'd ask.

"I'm so relieved that other people are taking care of him, and that you or I don't have to do it, that I'd pay twice as much for those darn diapers! You wish that I'd be more concerned about money, and I wish that you'd put it out of your mind. It's just another way Daddy is taking up space. He's like a big boulder sitting in the way!"

If I was light and loving—and could laugh about how differently we responded to the same situation, without trying to change or convince her—Rose lightened up, too.

She also responded well when I took her complaints to their own extreme, rather than trying to reason them away. Once I gravely suggested that she park herself with pen and paper in Archie's room twenty-four hours a day, in order to keep tabs on the actual numbers of diapers, wipes, and other incidentals. Although my mother opted not to devote her entire waking hours to this effort, she saw the humor in the suggestion, even as she gave it serious consideration.

Moving toward the Hot Issue

I never meant to forbid my mother to voice her pain or to muzzle her on any subject. While I used bantering and humor to deintensify her anxiety-driven focus on Archie, I also made it a point to move *toward* the very issues that brought her pain—but in a productive rather than a reactive way.

At calm times, I'd ask Rose questions to learn more about her experience. What was the hardest part about having Archie in the nursing home? Did she ever feel guilty that she wasn't able to care for him at home by herself? What was it like for her to have paid ten years of nursing home bills when she had saved and invested her money so carefully over the years? What were her worries about her own financial future? Did she think she would do better or worse after Archie died? Where did she think all that "worry energy" would go then?

I also made it a point to share my own problems with her, and to elicit her advice and perspective. Conversations such as these brought my mother and me closer together, rooted me deeper in my own history and identity, and allowed me to give my mother the attention and empathy she deserved.

DON'T RULE A SUBJECT OFF-LIMITS FOREVER

It's rarely helpful to rule a subject entirely out of bounds for all time. We may be tempted to say, "Mom, you can't talk about Dad anymore in my house!" Or, "I refuse to listen to how bad you feel because you're not doing anything to solve the problem!" But such a position will only offer short-term relief, if that.

Obviously, we need to define our limits ("Mom, I just can't listen to this now. I'm feeling too tense and preoccupied"). But to entirely forbid conversation about a hot issue drives it underground, which inevitably causes the other person's feelings about it to intensify. It leaves the proverbial elephant sitting in the middle of the room. And it leaves an already anxious person feeling more desperate because of being totally forbidden to voice compelling worries or complaints.

When family relationships are intense, it's far more useful to use humor, lightness, and imagination to deflect complaining and negativity. The tone of our voice is every bit as important as the content of our words. The challenge is to pass along less anxiety than we receive.

Say, for example, your mother is riveted on your dad in a negative way. No sooner is she off the plane than she corners you to exclaim, "Let me tell you what your irresponsible father did now!" It won't help to cross your arms in front of your chest and proclaim, "Mom, don't complain to me about Dad. Your problems with Dad are *not* my business. Please leave me *out* of this! You are putting me in the middle of a *triangle!*" Nor will it help to try to "reason" with your mother, join in her criticism, defend your father, or try to make her see the other person's point of view. It will be far more helpful to say something playful like, "Gosh, Mom, you've been married to that man for almost thirty years, and you still don't have him shaped up?"—and then to shift the topic to something else that will engage her.

If we're reactive to the level of repetition and negativity that a person brings to the conversation, we will tend to respond narrowly and habitually ourselves. Instead, we need to do the opposite and draw on our most creative self to help the conversation take a new and unexpected turn.

It's Another "Two-Step"

In addition to deflecting a conversation that's overloading us, we also need to return to it. Paradoxically, we can best defuse an anxiety-driven subject by moving *toward* that same subject, curiously and uncritically. Timing is an important factor, too. We're likely to get reactive if we're in the middle of preparing dinner and we pick up the phone, only to be immediately confronted with a family member's repetitive laments. It makes sense to get off the phone and then to reopen the discussion later when we feel solid and more centered.

Say, for example, that your dad is a serious hypochondriac. You may feel like the top of your head will fly off if you hear him complain one more time about his symptoms and his doctors, especially since he doesn't take good care of himself, and he doesn't take your advice, anyway. You can joke with him about his worries or shift the focus, because intensity on your part will only breed more intensity. You can also be inventive about approaching your father at a calm moment to learn more about the loaded issue of illness and doctors in the family. Over time, you might ask any number of questions that will give you a broader perspective. For example, you might ask, "Are there any other folks in our family tree who worried a lot about their health?" "Can you tell me more about Grandpa's stroke and how he coped with it?" "Do you think that your parents took good care of themselves, healthwise?"

It's totally counterintuitive to formulate a plan to open up a conversation on the very topic you want the other person to shut up about. If you can't stand your mother's anxious jabs at your single status, for example, you won't feel overly eager to approach her and say, "Mom, you seem worried about my future as a single woman. Can you tell me about your specific fears and concerns?" And, "Who are the single women in our family and how have they fared?" And "Mom, if you had never married, how do you think you would have managed in the world as a single woman?"

If we can ask questions that inspire thinking (rather than reactivity), we will elicit more empathic responses and respond more empathically ourselves. When we do voice our limits ("Mom, it's not helpful to me when you keep suggesting ways I can meet men"), the other person is more likely to hear us. But this higher level of conversation occurs best when we first make sure that others feel really heard and understood themselves.

"I JUST CAN'T LISTEN ANYMORE!"

Sometimes we have to draw a clear boundary to protect ourselves, especially if we are living under the same roof with the complaining or negative party. We need to say, "I can't live with this, and you need to get help." A client of mine, Gloria, took such a position with her intimate partner, Monique.

Gloria and Monique had lived together three years and hoped to be lifelong partners. Monique had a long-standing bent toward obsessive worry and self-doubt, but her ruminations intensified when work and family pressures combined to produce an especially stressful year.

Gloria felt sympathetic but got tired of listening to Monique's self-loathing comments about how she was a total failure and a deeply worthless person whose life was going nowhere. At first Gloria felt she had to keep listening, as if to do otherwise made her a heartless bitch. But later Gloria refused to indulge Monique. She used humor to encourage Monique to lighten up, or she'd tell her directly she didn't want to listen. Gloria would say, "Look, Monique, your brain is stuck in a negative groove, and it's driving me nuts. The more you keep thinking and talking this way, the deeper the groove gets. I think you need to kick your brain out of this groove and work on overcoming your negative thinking. I'm exhausted listening to you put yourself down."

Gloria learned to close her study door and tell Monique when she couldn't listen or be interrupted. Monique got sulky, and Gloria learned to let her sulk. No one has died from sulking, and a person can only sulk for so long. Gloria also spoke clearly and directly about the toll Monique's negativity took on their relationship and about the reality that Gloria might not stay in the relationship if nothing changed. She set limits in an overall climate of love and respect.

Finally, Gloria insisted that Monique get professional help with her problem, which Monique eventually did. Monique loved Gloria, valued her advice, and took Gloria's growing frustration seriously. Monique didn't want to blow the relationship by wearing Gloria down—and she was wearing herself down as well. Monique was able to make good use of therapy and medication, and she felt much better herself.

Because the emotional climate of their relationship was flexible and loving, and because they shared a common living space, Gloria had the clout to say, "You have to get help because your negativity is seriously jeopardizing our relationship"—and be heard. Relationships with parents and siblings tend to be less flexible, even when they are no less loving.

Whatever the relationship, we need to know the limits to our capacity for compassionate listening, and figure out how to protect ourselves when necessary. We also need to distinguish between a conversation in which the other person shares real pain, and a nonconversation in which chronic reactivity and negativity keep spilling over in our direction.

When the other person's brain is obsessively stuck on a particular topic, he or she may never change. But we'll do better ourselves if we navigate our part of the conversation in a solid and mature way. We'll also feel better if we have a strategy for shielding ourselves when necessary from the other person's intensity. Finally, connections are strengthened when we can both set limits and find a productive way back into the very conversation we're most allergic to.

CHAPTER 15

The Sounds of Silence: Finding a Voice When You're Rejected and Cut Off

Someone important to you may not be willing to speak at all. He or she may distance emotionally or even bolt from the scene. Maybe you need to talk, but the other person needs not to. The more history you've shared and the higher your expectations for the relationship, the more painful is the silence.

We like to believe that any crucial relationship can be repaired if only we persist in saying the right words. We want results. But in some circumstances your words will never touch the other person. You may have to let go of any expectation that the other person will ever change, or even allow you an opportunity to talk things through.

This challenge is hardest if we feel silenced and shut out through no fault of our own, and if we feel the depressed or angry emotions that normally accompany a terrible rejection. Finding a voice requires us to consider how we will speak and act (or not) in a relationship, based on our core values and beliefs—not simply *in reaction* to the other person's behavior or our own intense emotionality.

Rejection is not pleasant. No one intentionally signs up for it, or escapes it, for that matter. You may have a partner, a close friend, or a family member who lets you know it's over, then won't discuss it further. Let's look at all three situations, beginning with an intimate partner.

"HE WON'T DISCUSS IT!"

Mary sought my help a year after her fiancé, Bob, broke off their engagement but refused to discuss his decision. They had been dating almost two years and were about to buy a house together when Mary received the following e-mail:

> I've been doing a lot of thinking and I have to break off our relationship. My decision has nothing to do with you. You're a wonderful person and any man would be lucky to have you. I love you, but I'm not in love with you. I don't want to discuss it further. I'm very sorry for hurting you. You deserve better than me.
> Love, Bob.

Mary felt as if she'd been run over by a truck. She called Bob and asked him to talk with her about his change of heart. She told him that anything he could share with her was better than his silence, which fueled her own worst fantasies. Mary also pleaded with Bob to go with her to see a couples therapist, even for a single appointment. Bob agreed only to meet once in a café after work, during which time Mary cried and implored Bob to give their relationship more time.

Bob refused all Mary's requests for more conversation. He

stopped returning the calls she left on his answering machine, and he did not answer the letters she sent describing how devastated she felt and how horrible it was to end their relationship via e-mail. Still, she heard nothing. When his silence continued, she contacted Bob's only sibling and pressed him to provide her with more information, but she never learned anything more specific than "Maybe he got scared" and "Bob always had trouble with women."

"I Just Want to Understand!"

After Mary saw Bob at a concert with his arm around a new girl-friend, her already low spirits plummeted further, and she decided to call me. "I just want to understand *why* he left me," Mary explained. "If I can make sense of his behavior, then maybe I can forgive him and let it go." As it was, a year after their traumatic breakup, Mary was feeling worse rather than better, and she couldn't stop thinking about how Bob had treated her. "I just want to *understand,*" she said, over and over. She wanted to know how Bob could behave in a manner that was so cruel, callous, and out of touch—yet seem to suffer no remorse.

Abandonment is devastating. Mary experienced a profound loss and needed time to recover. Bob's sudden decision to bolt and the way he ended their engagement forced her to revise both her understanding of the past they shared and her picture of their future together. The breakup was particularly painful because she and Bob had never recognized or talked about problems as they occurred. Bob never discussed his doubts or shared any feelings of dissatisfaction, nor did Mary notice that something was obviously wrong. Naturally, she felt crushed and overwhelmed.

As more time passed, Mary's friends advised her, "Let it go!" and "Forgive and forget!" Perhaps they felt worn down by her repeated litany of complaints. But Mary just couldn't move on. She felt that Bob's actions were unforgivable, and it seemed as if she could prove her point only by continuing to suffer—not that this was her conscious intention.

Letting Go of Anger and Pain

Painful events happen to all of us, and we can become attached to our pain. We also get attached to the idea that if we stay angry long enough, and keep thinking about it hard enough, the person who wronged us will realize how terribly they've treated us—which won't ever happen, of course.

It's hard to give up the magical fantasy that hanging on to justified rage will someday force the other person to suddenly see the light and come groveling back to apologize and, most importantly, to feel equally if not more miserable. So let me say again that if it hasn't happened yet, the person who hurt you never will get it. And while you're sitting there ruminating about the terrible actions of your ex-girlfriend or ex-husband, that person may be out having a wonderful day at the beach. The fact that you're the only one suffering may be the best argument for stepping back from a negative attachment.

Did Mary have to forgive Bob? Of course not. Nor should she forget what he did to her, because what happened is a very real part of her history. This is not to deny the healing power of forgiveness. We know that forgiveness is divine and healthful and it certainly feels better to replace anger and hurt with love and compassion. Some people possess an extraordinary and overflowing capacity to forgive. Mary was not one of these people.

But she did need to make Bob less of a force in her life. Her anger kept her attached to Bob, even though her conscious intention was to get over him. Negative intensity preserves our sense of togetherness with the other person as surely as does positive intensity. Mary's anger was the glue that kept her stuck to Bob, although her anger was expressed by her ongoing obsession about "why" Bob had wronged her—which Bob probably didn't understand himself.

Mary needed to ask herself what was so special about this guy that she continued to give him so much power over her. Bob was probably never going to talk to her—yet two years after their breakup, Mary still hadn't moved on. Moving on doesn't mean forgetting or whitewashing the other person's behavior. It means protecting ourselves from the corrosive effects of staying stuck. Chronic anger and bitterness dissipate our energy and sap our creativity. Each of us has a certain amount of energy that fuels our spirit. If 5 percent—or 75 percent—of that energy is directed toward hating someone who's wronged us, then that same percentage is not available for other pursuits. And being overfocused on any one loss can cause us to *under*focus on other losses, betrayals, and deceptions from our past that also merit our attention.

Sometimes we can do nothing to force the other person to talk to us. The challenge for Mary was to allow her negative feelings to recede, so that she could feel more loving, powerful, peaceful, and whole. If her anger and pain kept her stuck in the past, she couldn't live fully in the present, nor could she move forward into the future with optimism and hope. For reasons of his own, Bob was not able to honor her or be kind to her. So she needed to honor and protect herself by refusing to let painful emotions loom so large in her day-to-day life.

Two Years after the Breakup

Communication with Bob did open up a bit after Mary began putting her energy into living her own life as well as possible. She was excited about a new job, was taking good care of herself, and was no longer focused primarily on Bob. Earlier, Bob had sensed Mary's desperation and walled himself off entirely. Perhaps he didn't want to feel worse than he already did, and he may have been unaware about what propelled him out of the relationship. He may also have feared that talking to Mary when he was leaving her might weaken his resolve.

Bob didn't show maturity or kindness when he cut himself off from Mary at the time of the breakup. One wishes for his sake, as well as hers, that he would have behaved more honorably. Bob's experience of himself probably suffered when he left by the easiest and least courageous route. But as long as Mary's spoken or unspoken attitude was, "My life is so damaged, all I want to do is talk to you so I can understand what happened," Bob wasn't going to put himself in an emotionally vulnerable position. He probably also knew that nothing short of going back to Mary and working on their relationship would truly assuage her pain. Mary was saying, "I just want to *understand*," but that wasn't exactly true, or at least not the whole story. She also wanted him to suffer the way he had made her suffer—a perfectly normal human impulse.

In Praise of Brevity

With my help, Mary eventually composed a brief letter to Bob, telling him about her new job and sharing a few thoughts about her part in the difficulties between them, such as the fact that she must have been sleepwalking not to have been aware of any problem. She told him that she'd appreciate hearing his perspective on their relationship, its strengths and limitations, from his current vantage point, including any thoughts he might have about the breakup. Mary ended the letter by saying, "If I don't hear from you, I wish you well in all your ventures—and give Poochie [Bob's dog] a big hug from me." She started and ended the letter on a light note.

Mary revised this letter a few times before she mailed it. I assured her that the original four-page, single-spaced typed document she was planning to send would repel Bob like a punch in the nose. Indeed, if you want to ensure that someone *won't* hear you, just write a really long, detailed letter, maybe even proving your point in different ways, just in case the person didn't get it the first few times around. When the other person (be it your lover, ex-husband, mother, or child) is already in an avoidant mode or heavily defended, you have to lower the intensity—not raise it—if you hope to begin a conversation.

The Other Person May Be Clueless

Although Mary heard nothing from Bob, she was able to understand his silence as simply a sign of how uncomfortable he felt about the termination of their relationship. Several months later, she bumped into him at a shopping mall and invited him to sit down with her over coffee. Bob accepted, which I believe had everything to do with the tone of Mary's earlier letter and her ability to greet him cordially. After some small talk, Bob took the initiative to say, "I'm sorry I broke up with you the way I did. I just couldn't see the point of doing it in person because I didn't know what to say." When Mary asked him more about his decision, he said, "I needed to get out. I felt it was a mistake for us to get married. I really don't have a good explanation."

Persons who initiate breakups may truly not know why they do what they do. Their behavior has more to do with them than with us, but that doesn't mean they can unravel their own motives. As much as we crave a satisfying explanation, the other person may never admit the truth—or even know it. Humans have a remarkable capacity for self-deception, and obviously Bob couldn't tell Mary what he didn't know himself. So Mary got very little information to satisfy her need to make sense of why Bob left her.

To her credit, Mary was able to share a bit of her own perspective without attacking Bob, but also without letting him off the hook. She told him, "I know I was in denial not to notice any problems in our relationship. And I understand that my own intensity made it more difficult for you to talk to me about leaving. But to be honest, I'm still struggling with some leftover anxiety about what happened. Your refusal to talk with me made it difficult for me to understand what happened and to get some closure and peace of mind." Bob said, "I'm sorry," and Mary said, "I appreciate that." She left the conversation feeling good about how she had handled herself.

Later Mary told me that the hardest part of the whole ordeal was her belief that Bob didn't really feel sorry at all. "I'm innocent, and I'm the one who's suffering. He's a schmuck who slips into a new relationship without looking back. I'd like to think that what goes around comes around. But I'm not sure."

We can never know for certain what price Bob paid for his behavior or how it will ultimately affect his identity and self-esteem. I believe that whenever we diminish and disregard another person, we also diminish our own self. But who knows whether Bob will be punished in this life or the next, as Mary secretly wished? More important, Mary had stopped holding onto her pain as a way to prove what damage Bob had done.

Making Sense of His Behavior

Therapy provided Mary with a neutral space where she could piece together her own understanding of the breakup. I learned that Mary had first met Bob eight months after his mother was killed when her car was struck by a train. In the anxious emotional field of his loss, he had tenaciously attached himself to Mary. Bob's mother had been a severely depressed woman whose death left him with wrenching questions, including his fantasy that the accident might really have been a suicide. His father had died from a heart problem when Bob was sixteen, and losing his mother surely revived that earlier loss as well.

Perhaps there was a connection between being on the receiving end of this devastating loss and Bob's sudden and inexplicable decision to skip out on Mary. Perhaps he was creating in Mary the same helplessness that he himself had felt on losing his mother. His mother's abrupt departure left him with a lot of unanswered questions, which is exactly where he left Mary later on.

For Bob, there might have been many other factors fueling the beginning and the end of this relationship. We can never know for sure what "causes" another person's behavior; it's hard enough to know this for ourselves. Still, Mary found it helpful to put Bob's behavior into a broader context and not take their breakup so personally. We all need to step back from any problem and view it through a wide-angle lens.

Finally, Mary began to look squarely at her own part in the loss she didn't see coming. She and Bob had gotten off to a quick start with steamy sex but had never slowed down enough to get to know each other better before getting engaged. She had fallen madly in love with him and thought that love was enough to carry their relationship forward. While it's unwise to marry someone we don't love or feel attracted to, love alone is never a good enough reason to marry. Ditto for sex. As Garrison Keillor puts it, "Marrying for sex is like flying to London for the free peanuts and pretzels. It's not the point of the thing, is it?"

Mary learned the hard way how important it is to take time to really get to know each other. It's our job to be as clear-eyed and alert as possible, to do as much talking and listening and observing right up front. Even then, there's no guarantee the other person won't surprise us in an unexpected and painful way, whether in love or friendship. We can't make another person talk to us or own up to the truth as we see it. As we've seen earlier, the greater the misdeed, the less likely a genuine apology will be forthcoming, especially if the harmed party is trying to impress the wrongdoer with the damage he's done. We can, however, stay as aware as possible and try to live our own lives as well as possible. Mary was ultimately able to do this.

There may be more than one time in your life when a key person will not engage you in real conversation or own up to bad behavior. It may be your lover, an ex-spouse, or a once-dear friend. The other person may literally refuse to talk to you, as Bob did with Mary. Or (as the next story illustrates) she will show up for the conversation but not be present, emotionally speaking. Nothing you say can touch the other person, because she has already taken herself out of the relationship and will not allow herself to be moved or empathically affected by what you say. Her need to maintain a sense of goodness and righteousness, and her intolerance for feeling guilty and vulnerable, makes such a person unavailable for a genuine and authentic exchange. Nothing you can do or say will ever really reach her.

So we need to step aside from the pain and move on, which requires us to give up the fantasy that the other person will feel bad about what's been done or will see things as we would like. When Bob left Mary, there was nothing she could do to bring him back, even to talk. She needed instead to get all the love and support she could find elsewhere to keep breathing and to stay ambulatory through a very difficult situation.

FRIENDSHIP MATTERS

A greeting card depicts a young woman telling her female friend, "I'll always be there for you." Inside the card is the postscript, "Unless, of course, I have a date." It reminds me of the old stereotype that a girlfriend is something to do until a man comes along. But that way of operating doesn't hold up.

Generally speaking, friendship is what women do best. A solid friendship can survive anger, jealousy, competition, and the rest of the entire range of human emotions—especially if we can recognize these feelings rather than deny them ourselves. But friendships don't always go easily or well, and we can feel deeply hurt or disappointed by a friend at any stage in life. When tension or distance persists in a close friendship, it makes sense to find some way to address it. But as the following story illustrates, that's not always possible.

WHEN A FRIENDSHIP FALLS APART

Corrine and Joan became inseparable best friends when they were college freshmen. Their friendship had always been one-sided, with Joan the leader, the teacher, the in-charge person who did the giving, nurturing, and planning. Corrine described herself as more of a follower, who had difficulty knowing her own ideas or tastes and who readily deferred to Joan. Despite this lack of balance in their friendship, they were each other's closest confidantes. Corrine loved Joan, but after Corrine married, their friendship fell apart. Corrine reported that first Joan distanced, then acted almost repelled by her.

At the time Corrine sought my help, she had already made a number of important efforts to reconnect with Joan. First she initiated activities, such as calling Joan to play racquetball or shop for clothes. In this way, she let Joan know that their friendship was still important to her. Joan accepted some of the invitations, but never reciprocated. An unmistakable feeling of disconnection and superficiality settled between them, hanging in the air like thick fog.

Corrine also told Joan directly that she had noticed a big rift in their friendship. She asked whether Joan shared her observation and what Joan's perspective was on how their friendship had changed since Corrine's marriage. She talked to Joan without criticism or blame and without pressuring or pursuing her for closeness. In response, Joan kept insisting that she was just very busy and there was nothing to discuss. Corrine let some time pass before broaching the subject again. She said, "Joan, I *know* that something is wrong. I'm wondering if I did something in the relationship that upset you. It's important for me to know if I've offended you or upset you in some way."

Each time Corrine tried to talk about the distance and tension between them, she felt that Joan disqualified her perception of reality. Corrine left every conversation feeling diminished and down. She would then let some time pass before going back for another round. In a final conversation, Corrine was very firm. She asked Joan if she could come over and talk to her for ten minutes. Sitting in Joan's kitchen, she said, "Joan, it's clear to me that something is very wrong. You keep telling me that there's nothing to talk about, but I can't accept that. We were close friends for six years. I'd be denying my whole sense of reality if I tried to convince myself that we just drifted apart because of our busy lives. It's clear to me something is going on. I'm not asking you to change how you think or feel. I'm just asking you to help me make sense of what's happened, including my part of it."

Joan sat like stone and said, "I'm sorry if your feelings are hurt." Corrine's attempts to take the conversation further went nowhere. Corrine is a person with a deep need to talk things through, and her natural tendency is to keep hammering away. To her credit, she let this response from Joan be the final round.

Corrine suffered an important loss, and her anger and pain lasted a long time. The fact that Joan was unable to talk about what happened in a real way only increased Corrine's difficulty in finding peace of mind. In addition, every loss evokes feelings from prior losses and betrayals, and Corrine later recognized she was grieving for more than her lost friendship with Joan. And when we depend on one friend to meet all our friendship needs—as Corrine did with Joan—we place ourselves in a very vulnerable position.

Letting Go

The challenge Corrine faced was identical to the challenge Mary faced when Bob split with her. Both needed to let go of the fantasy that the other person would one day talk to her or "see the truth," that the old friendship/romance could somehow be restored, or that they would ever understand why the other person jumped ship. Corrine could do nothing to change or fix the situation. She will never understand Joan's transformation, so she must cope with the difficult challenge of ambiguous loss.

We never know for sure what motivates other people. Still, the human desire to construct explanations for other people's behavior is very strong. Perhaps Joan derived a profound satisfaction from nurturing and taking care of Corrine—a satisfaction that tapped into her own needs. When Corrine married, Joan may have felt betrayed (and out of a job) because Corrine now had another person to care for her and be more central in her life. If this is so, Joan will probably never be aware of her feelings. Certainly, it wouldn't be useful for Corrine to interpret Joan's behavior to her, as if she were Joan's therapist or had the answers. Joan may have been propelled out of the friendship by something that will remain a mystery to both women.

Corrine needed to keep in mind that the sad ending of this important friendship did not negate the fact that she and Joan had cared deeply for each other, and that they both gained something important during their six years of camaraderie. But the healing conversation Corrine hopes for probably won't ever take place. Although it takes two people to form an intimate relationship, it takes only one to end it.

Learning to Leave the Table

Corrine's efforts to engage Joan in conversation made sense because something had propelled Joan into profound emotional distance. In other cases, relationships simply go through cycles of closeness and distance, and we don't have to try to process everything. A good friend may simply drift away, lose interest, or find other friends to hang out with. Here, women could take a lesson from men, who can appreciate that not everything has to be addressed. When a friend is no longer as invested in the relationship as we are, we can learn to feel more at home with a distant or outside position.

Although it can be painful, sometimes a person we love seeks distance that we have to accept. Our friends are free to be friends with whomever they choose. Their feelings for us may wax and wane. We want everyone we love to be loyal and stable figures in our lives, but we can't always have that. Change and impermanence are part of every relationship, and we can't hold the clock still, much as we may try.

None of us can escape rejection and disappointment unless we sit mute in a corner and take no risks. If we live courageously, we will experience—and survive—rejections and losses that are not fair and not talked through. Sometimes in our lives the best course of action is to let go and move on. I wish I could give six easy steps to accomplish this goal with ease, grace, and efficiency. But it really does take time. We can't just decide that letting go is the mature and reasonable thing to do, and then, "Poof!" make it happen.

My college friend Ralph tacked a poster on his kitchen wall that said:

YOU'VE GOT TO LEARN TO LEAVE THE TABLE WHEN LOVE'S NO LONGER BEING SERVED

Ralph adored this poster, and so did I. I can still picture him sitting in the student union in Madison, Wisconsin, chanting those words like a mantra in his deep, sonorous voice and with great dramatic flair. Sometimes what we have to do is as simple—and as difficult—as that.

A caveat: My advice is different if you've been cut off by an important family member. Say you're not welcome in your mother's home or your adult daughter isn't speaking to you. If that's the case, the solution is *not* to leave the table.

Family therapist Monica McGoldrick notes that when people marry, they often make a vow in front of God and everybody to stay connected till "death do us part." But we know that at least half of these marriages will end in divorce. Monica suggests that we might be wiser to make this vow instead to our parents and children. Marriage lasts for as long as it lasts, but we can't orphan ourselves from our first family, even if we try.

In families, distance and cutoff don't imply a lack of feeling, as we commonly assume. ("What kind of mother could abandon her child like that!" or "How heartless of him to neglect his mother that way!") Rather, distance and cutoff speak to an intensity of emotion that makes contact too difficult.

"MY SON WON'T SPEAK TO ME!"

How do you *not* leave the table when a key family member has cut you off? What follows is a long story made short.

Jackie came to see me because her son, Gregory, a successful engineer in his twenties, wouldn't speak to her. A year earlier Jackie had divorced Gregory's dad and married her supervisor at work. Gregory saw his dad as the victimized, done-in partner, and he planted himself firmly in his father's camp. Refusing all contact with his mother and her new husband, he said, "You and that man are never welcome in my house." And, "As far as I'm concerned, I don't have a mother."

When I met with Jackie, she had already made a number of attempts to restore her connection with her son. She had tried to "reason" with him (which boiled down to trying to convince him to see things her way). She sent him a letter explaining his dad's contribution to the problems in their marriage, which only intensified Greg's loyalty to him. She showed up on her son's doorstep, then cried on his front porch when he wouldn't let her in. Finally, Jackie's husband jumped on the bandwagon and wrote Gregory a letter telling him how hurt his mom was and asking him to please allow her back into his life.

These efforts, quite predictably, made things worse. Jackie then concluded that if Gregory was going to deny her existence, she had no option but to erase him from her life.

Walking a Fine Line

Jackie did need to accept the fact that her son might never speak to her again. But she discovered that she didn't have to erase him from her life. As we've seen, finding a voice requires us to speak and act in accordance with our core values and beliefs—not from a place of emotional reactivity. An important family member may choose to treat us as if we don't exist. But if it's not congruent with our values to respond in kind, we shouldn't. We need to define our differences.

Gregory's position was, "I don't have a mother." Jackie's position was, "I can't pretend I don't have a son." This difference required Jackie to walk a very fine line. How does one stand for connection, while respecting the position of distance taken by the other party? With the help of therapy, Jackie was able to chart her course.

Jackie decided to send an occasional card, without requesting or expecting a response, and without trying to change Gregory's mind in any way. In the first card, she wrote a short note acknowledging that her decision to divorce and remarry had caused him a great deal of pain. She said she understood that he needed space, and she didn't want to be intrusive. She explained that she was sending a card because it was too painful for her to deny the existence of her own son—she couldn't pretend that he didn't exist.

Jackie continued to send cards on holidays and birthdays and to leave a brief phone message on the machine when there was important family news. ("I don't know if you heard that Uncle Ed is having heart surgery next week.") She reduced her expectations to zero in terms of getting a response. She also kept her communications brief, low-key, and well-spaced over time. To do otherwise would be to disregard the boundaries Gregory had set. In one birthday card, she told Gregory that she loved him and that if he ever wanted to talk or just meet briefly for a cup of coffee at any time in the future, she would like to do that. But she didn't try to change or convince him. Jackie also began working on relationships in her own family of origin that were distant and cut off, and she put her primary focus there.

Two years into therapy, Jackie heard from her sister that Gregory was getting married. She sent a card congratulating him, then later sent a modest and thoughtful wedding gift. But it wasn't until after Gregory and his wife, Marlene, adopted a baby boy that a crack appeared in the ice. Marlene called to thank Jackie for her baby gift, and to invite her to drop by the house on a weekday afternoon to see the baby. Jackie knew that Gregory would be at work, but graciously accepted the invitation without suggesting a different time. During this first meeting, both women were cordial and pleasant. Jackie talked about how adorable the baby was and resisted the temptation to talk about how hurt she was by Gregory's rejection.

Jackie and Gregory connected for the first time almost four years after Jackie called me to initiate therapy. Gregory had slowly softened and began to include his mother in his new family. Their conversations focused on the baby, a neutral and delightful subject around which they could interact without intensity. But even if Gregory had never spoken to his mother again, Jackie was not going to be locked into total silence. Rather, she used her best thinking and intuition to find a balance between staying in touch and respecting her son's wish for no contact. Obviously, Gregory could choose whether to read her cards, stick them unopened in a drawer, trash them, or return them unopened. One hopes for his sake that he will forgive his mother for the choices she made and that he will choose to include her in his new family. Such forgiveness is the very best gift he can give to himself and his son.

When it comes to family cutoffs involving parents, grandparents, children, and siblings, I try to help people to be enormously patient for the rest of their lives. You can express the wish for connection (which is not the same as rescuing or bailing out an irresponsible family member) and stay in some contact. You can keep your heart open to the possibility that a conversation—even if it's only about the weather—may one day occur. A reconciliation may never happen, but what's important is where you stand.

To Thine Own Self Be True

No matter how clear or creative our voice, or how persistent our efforts to be heard, we may not get the results we want. Other people may fail to respond to us with sensitivity, empathy, or fairness. They may refuse to respect our request for even a simple behavioral change. They may not show up for the conversation. Still, we may choose speech over silence as a way to honor our own need for language and self-definition.

When I was growing up my mother quoted Shakespeare: "This above all: To thine own self be true." While Shakespeare used these words ironically, they are good words to live by. If we are untrue to ourselves, we live disconnected lives, and we cannot love ourselves or anybody else very well. But as we've seen, having a true voice isn't always the same as spontaneous, unvarnished candor. I was reminded of this fact when my friend Jeffrey Ann called the other day and said, "Harriet, I felt picked on during our walk yesterday. I felt like you were on my case and focusing on my weaknesses.

Maybe some of what you said was true, but it didn't feel good." She was absolutely right. Nowhere in our conversation had she asked for my feedback. Who appointed me the expert on her problems? And what good are my frank observations, right or wrong, if they left her feeling diminished? Of course, I apologized. I had also left the conversation feeling a little bit down, so my body had registered that something wasn't right.

Along with a spontaneous voice, we need restraint as well as the courage to experiment with novel or unfamiliar parts of ourselves. Sometimes there is a gap between what we say and what we really feel. This isn't necessarily a problem. Sharing "true feelings," while essential in certain circumstances, is highly overrated as a principle to live by. But sometimes there is a gap between what we say and what we truly hope to accomplish in a relationship. Or a gap between what we say and the sort of person we hope to become. Or a gap between what we say, and a deeply held value, belief, and principle. And *that's* a problem.

Having an authentic voice is not about speaking from a place of angry reactivity, righteousness, or criticism. Rather, it's about constructing a more solid and whole self, modeling the behavior we want from others, and thinking about relationship problems with clarity, creativity, and wisdom. Steps in this direction require us not to rush in and tell all, but rather to consider how our words affect others.

Finally, our clarity of voice reflects the degree of our self-awareness. Having an authentic voice requires us to operate from core values, rather than in reaction to the other person's immaturity. We must keep our own immaturity in check, which admittedly is hard to do when we're caught up in strong emotions. We need to use both wisdom and intuition in deciding whether to lighten up and let something go, or to take a difficult conversation another round.

LISTENING IS CONNECTING

Having an authentic voice requires us not only to speak wisely but also to listen well. When we listen to another person with attention and care, we validate and deepen the connection between us.

We've seen how hard it is to be fully emotionally present, without defensiveness or distraction, when confronted with the other person's anger, criticism, suffering, or just plain kvetching. We need to define the limits of our capacity to listen and refuse to engage in hurtful or downward-spiraling conversations. But many of us have trouble simply *being* with another person, much less listening with our full presence.

A Life-Altering Dinner Party

For example, my New York friend Audrey, a gifted musician and cook, is also a wonderful storyteller. The only child of doting parents, she was rewarded, if not glorified, for her talent with words, her precocity, the way she could entertain and impress her parents' friends and other adults. She carried this behavior unmodified into adulthood, especially in group situations. When another person would tell a story, she often grabbed the first empty space by saying, "That reminds me of something that happened to me." Then she would proceed to tell an even more dramatic and longer story about her incredible trip to Paris or her near-death experience hailing a cab in Chicago.

Audrey was admired by her friends for being such an interesting person, which made it harder for her to observe the absence of mutuality in her connections. People were impressed by her, but didn't leave her company feeling a greater sense of worth themselves. Audrey had no idea that the gift of listening could be an even greater gift than her ability to speak brilliantly.

A turning point came during one of her small, elegant dinner parties. A guest I hardly knew named Stanley was asked about his daughter's recent diagnosis of ovarian cancer. Not too long after he began speaking, Audrey said, "I know exactly what you mean," and turned the conversation to her own experience with a serious health scare when she was in college. She had spoken for only a minute or two when Stanley suddenly began sobbing. He collected himself at once, and offered a simple apology to his host. "I'm sorry, Audrey," he said softly. "I didn't mean to interrupt. I just can't listen right now because I'm too preoccupied with my daughter."

Of course, Stanley had no reason to apologize, and Audrey knew that. She felt ashamed of her insensitivity. This painful event led her to a new level of insight and capacity for self-observation. Audrey told me that Stanley's tears made her feel like the Wizard of Oz when Toto pulled the curtain off track and exposed a big phony. But what Stanley had exposed, really, was the fact that Audrey was an ordinary, flawed, uncertain human being, like the rest of us.

Audrey decided to practice listening. She was disciplined in all her pursuits: cooking, playing the flute and cello. Now she chose to apply that same discipline to listening. At her next dinner party, she experimented with only listening and asking questions. When asked about herself, she made an effort to answer simply, without elaboration and her usual superlatives. Practice is everything, whether we're aiming to take up more—or less—space.

Obviously, Audrey didn't change her habitual, reflexive habits with one gigantic act of will. But she kept practicing. Over time, this shift to listening allowed her not only to know others better but also to be known. She moved in the direction of greater self-acceptance and a more balanced and accurate picture of herself. Audrey discovered that being "ordinary" wasn't a terrible trait but rather a centering human experience. When she spoke, she did so with more thought and consideration. This change, which initially required "watchful effort" on Audrey's part, eventually led to her feeling more relaxed and more herself. Of course, how we listen to the people we love is obviously more important than how we listen in a group or social engagement.

Listening Is Being *and* Doing

As Ram Dass has pointed out, we are human *be*ings, not human *do*ings. *Be*ing is very hard for some of us, and we may need to rehearse silence more than we need to practice speech.

But empathic listening is active. First, we may need to invite the other person to tell us stories and experiences. When people suffer, they often suffer twice, first because they have lived through something painful, and second, because family members or close friends either don't want to hear about it or don't communicate a wish to hear about *all* of it. I continually observe people who love each other getting locked into terribly lonely positions. A parent, for example, doesn't want to be "intrusive" by asking a son or daughter about something painful or sensitive, and the son or daughter doesn't want to "burden" the parent with painful facts that appear not to be wanted. Nothing takes a greater toll than a "Don't ask, don't tell" policy in family life when the subject is emotionally important.

No one wants to be intrusive or pushy, but it's better to err in the direction of truth, of offering the invitation. Cara, a woman I saw in therapy, had a twenty-five-year-old daughter, Shawn, who was extremely private. Shawn had been raped in college, and her parents had offered emotional support but had never asked about the details of the rape and its aftermath. I encouraged Cara to reopen the subject with her daughter, to express her wish to know what took place, and her sorrow that she had distanced from the details at the time of the trauma. When she did so, Shawn responded abruptly with, "I don't want to talk about it." The next day Cara called her and said, "Shawn, I totally understand your not wanting to talk about what happened. I don't want to push you or be intrusive. But I want you to know that if you are ever ready to talk about it, I'm here to listen."

Cara then brought the subject up with care and sensitivity at several future occasions when an opening presented itself. Cara knew that she had played an important role in solidifying her daughter's silence about the rape, by not inquiring about the facts when it had happened, and by not mentioning the subject in the five years that followed. So it was not sufficient for Cara to offer just one green light and then withdraw again. In time Shawn did respond to her mother's invitation, and their relationship became much closer.

Whatever words we use, we can extend an invitation to the people we love to tell us their stories when they are ready. If a person we love has found the courage to live through something, then we can find the courage to listen, to give our full attention, and to not back away from asking, "Is there more you haven't told me?"

"Under Construction"

The challenge of conversation is not just in being our self but in choosing our self, since what we call the self is constantly reinvented through interactions with others. The self is always under construction.

Paradoxically, the more enduring a connection, the more vulnerable we are to getting stuck in conversations where our experience of our self and the other person becomes fixed and small. Disconnection can become a way of life for people sharing the same home, a common history, or the same bed. In couple relationships and family life we may need to make a special effort to engage in novel conversations that will create a larger view of who we are and what our relationships can become.

Our conversations invent us. Through our speech and our silence, we become smaller or larger selves. Through our speech and our silence, we diminish or enhance the other person, and we narrow or expand the possibilities between us. How we use our voice determines the quality of our relationships, who we are in the world, and what the world can be and might become. Clearly, a lot is at stake here.

Notes

Prologue: BACK TO THE SANDBOX

p. xiii.   I heard the sandbox story from Janis Abrahms Spring, author of *After the Affair: Healing the Pain and Rebuilding Trust When a Partner Has Been Unfaithful* (HarperCollins, 1997) who credits Rabbi Israel Stein.

Chapter 1: FINDING YOUR VOICE

p. 5.   "we are trying": Adrienne Rich, "Women and Honor: Some Notes on Lying," in *On Lies, Secrets, and Silence: Selected Prose, 1966–1978*, rev. ed. (New York: W. W. Norton & Company, 1995).

Chapter 2: VOICE LESSONS FROM MY FATHER

p. 22.   "Sometimes I think we would lose ourselves": Kat Duff, *The Alchemy of Illness* (New York: Crown, 1994).

Chapter 3: OUR FIRST FAMILY: WHERE WE LEARNED (NOT) TO SPEAK

p. 26.   the family is a sensitive system: Thanks to Betty Carter and Monica McGoldrick for their work on predictable and unpredictable stages and stresses of the life cycle. See Carter and McGoldrick, eds., *The Expanded Family Cycle: Individual, Family, and Social Perspectives*, 3d ed. (Needham Height, Mass.: Allyn & Bacon, 1999).

p. 30.   Our cultural heritage: Monica McGoldrick, "Ethnicity and Family Therapy: An Overview" in M. McGoldrick, J. Giordano, and J. K. Pearce, eds., *Ethnicity and Family Therapy*, 2d edition (New York: Guilford Press, 1996).

Chapter 4: SHOULD YOU SHARE VULNERABILITY?

p. 40.   "Years ago, my husband": Carol Tavris, "How Friendship Was Feminized," *New York Times*, May 28, 1997.

p. 47.   "We have this notion": "Between Living and Dying: A Conversation with Anne Finger about Abortion and Assisted Suicide," *Sun* (Chapel Hill, N.C.), no. 252 (December 1996).

p. 47.   "Like a stain on our clothes": bell hooks, *All about Love: New Visions* (New York: William Morrow & Co., 2000).

p. 47.   "consumer attitude toward experience": Michael Ventura, "A Primer on Death," in *Family Therapy Networker*, Jan/Feb 1996

Chapter 7: LOVE CAN MAKE YOU STUPID

p. 98.   "I nail his underwear to the floor": Quoted in Carolyn G. Heilbrun, *The Last Gift of Time: Life beyond Sixty* (New York: Ballantine Books, 1998).

Chapter 10: WARMING THINGS UP

p. 145.   When I work with couples: Ellen Wachtel, *"We Love Each Other, But . . .": Simple Secrets to Strengthen Your Relationship and Make Love Last* (New York: St. Martin's Griffin, 1999); Betty Carter and Joan Peters, *Love, Honor, and Negotiate: Making Your Marriage Work* (New York: Pocket Books, 1997).

p. 149.   "the four horsemen of the apocalypse": John Mordechai Gottman and Nan Silver, *The Seven Principles for Making Marriage Work* (New York: Three Rivers Press, 2000).

p. 151.   Ellen Wachtel's STOP rule: Wachtel, *"We Love Each Other, But. . ."*

Chapter 11: SILENT MEN/ANGRY WOMEN

p. 158.   how triangles operate: My work with stepfamilies is grounded in the theoretical contributions of Monica McGoldrick and Betty Carter.

Chapter 13: AN APOLOGY: DON'T HOLD YOUR BREATH

p. 196.   On the issue of holding men accountable and "inviting the wrongdoer to accept responsibility," I am grateful to conversations with narrative therapist Julie Cisz, the writing of Alan Jenkins, including his *Invitations to Responsibility: The Therapeutic Engagement of Men Who Are Violent and Abusive* (Adelaide, Australia: Dulwich Centre Publications, 1990, and the work of Rhea Almeida and her colleagues. See Rhea Almeida, Rosemary Woods, Theresa Messineo, and Roberto Font, "The Cultural Context Model," in *Re-Visioning Family Therapy: Race, Culture and Gender in Clinical Practice,* ed. Monica McGoldrick (New York: Guilford Press, 1998).

Chapter 14: COMPLAINING AND NEGATIVITY: WHEN YOU CAN'T LISTEN ANOTHER MOMENT

p. 201.   "Oy, am I toisty": Leo Rosten, *The Joys of Yiddish* (New York: Pocket Books, 2000).

Author's Note on Professional Acknowledgments

Books are the products of many people's ideas. There is no way to adequately acknowledge all of one's intellectual debts, especially when an author does not limit her subject but rather wanders over a vast terrain. For each person I thank by name, I am aware of countless others also deserving of my thanks. So I've chosen to use broad brush strokes and acknowledge categories of thinkers who have influenced my writing and clinical work. My intent is not to acknowledge only their specific contributions, but also to symbolically honor the many women and men I've learned from.

In this spirit, I want to thank:

- All the feminist thinkers who have so passionately and uncompromisingly told the truth and incisively challenged the dominant beliefs about what's true and whose truth counts.

- Murray Bowen and the family therapists (including Katherine Kent) who taught his theory and brought it home for me. All my books, including this one, draw on key principles of Bowen Family Systems Theory.

■ Women's Project in Family Therapy (Betty Carter, Peggy Papp, Olga Silverstein, and Marianne Walters) and Monica McGoldrick, for their pioneering, inclusive, creative work on families and their tireless teaching.

■ Jean Baker Miller and the Stone Center, Wellesley, Massachusetts (including Irene Stiver, Judith Jordan, and Janet Surrey, among others), for their innovative relational approach to female development and their theoretical contributions to the subject of connection.

■ Michael White, David Epston, Alan Jenkins, Johnella Bird and Kathy Weingarten, for narrative work (a new theoretical perspective for me), which is radical in its commitment to facilitating respectful and enlarging conversations between therapist and client.

■ Carol Gilligan, Deborah Tannen, Mary Pipher, and other authors who have brought the academic subject of female voice into mainstream consciousness. Thanks also to Ron Taffel, Stephen Bergman (a.k.a. Samuel Shem), and Terrence Real, among others, who have written about men's struggle with connection, and to Edward Hallowell for his book *Connect*.

■ The Menninger Clinic, for being my professional home since I began a postdoctoral fellowship there in clinical psychology in 1972. As I finish this book, I am terribly saddened that Menninger is closing shop in Topeka.

■ Emily Kofron, for being *both* professional colleague and the friend whose love, support, advice, and aesthetic, political, and clinical sensibilities I totally rely on. May we always live within walking distance of each other.

INDEX

BOOKS BY HARRIET LERNER, PH.D.

THE DANCE OF FEAR
Rising Above Anxiety, Fear, and
Shame to Be Your Best and Bravest Self

ISBN 0-06-008158-9 (paperback)

Using her wonderfully rich and inviting therapeutic voice,
Lerner teaches us how to cope with fear, anxiety, and shame
in order to live a loving, happy, and creative life.

Previously published as *Fear and Other Uninvited Guests*
in hardcover.

THE DANCE OF ANGER
A Woman's Guide To Changing the
Patterns of Intimate Relationships

ISBN 0-06-074104-X (paperback)
ISBN 0-06-072650-4 (abridged CD)
ISBN 0-89-845796-3 (abridged cassette)

In this acclaimed classic, Lerner teaches us how to identify
the true sources of anger and to use it as a vehicle for
creating meaningful and lasting change.

THE DANCE OF CONNECTION
How to Talk to Someone When You're Mad, Hurt, Scared,
Frustrated, Insulted, Betrayed, or Desperate

ISBN 0-06-095616-X (paperback)
ISBN 0-69-452545-6 (abridged cassette)

This life-changing book teaches men and women bold new
"voice lessons." We learn how to heal betrayals and broken
connections, and to speak with clarity and strength, even
when the other person behaves badly.

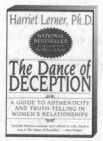

THE DANCE OF DECEPTION
A Guide to Authenticity and Truth-Telling
in Women's Relationships

ISBN 0-06-092463-2 (paperback)
ISBN 0-06-072664-4 (abridged CD)

From family secrets to sexual faking, this illuminating book
about lies and truth-telling shows women how to build
authentic relationships based on trust.

Visit www.harrietlerner.com

THE DANCE OF INTIMACY
A Woman's Guide to Courageous Acts of Change in Key Relationships

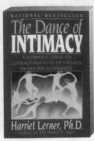

ISBN 0-06-091646-X (paperback)
ISBN 0-06-072654-7 (abridged CD)
ISBN 1-55-994147-2 (abridged cassette)

With wisdom and warmth, Dr. Lerner teaches the specific steps necessary to heal relationships challenged by too much distance, intensity, or pain.

LIFE PRESERVERS
Good Advice When You Need It Most

ISBN 0-06-092835-2 (paperback)

Illuminating answers to women's most-frequently-asked questions. Dr. Lerner covers the landscape of work and creativity, anger and intimacy, friendship and marriage, children and parents, loss and betrayal, sexuality and health, and much more.

THE MOTHER DANCE
How Children Change Your Life

ISBN 0-06-093025-X (paperback)

A witty, touching, and unconventional look at mothering, and how it transforms us—and all our relationships—inside and out.

WOMEN IN THERAPY

ISBN 0-06-097228-9 (paperback)

A rich store of professional wisdom about the problems women bring to therapy—anger, fear, depression, dependency, guilt, self-sacrifice, and self-betrayal.

CHILDREN'S STORIES FROM HARRIET LERNER AND SUSAN GOLDHOR

WHAT'S SO TERRIBLE ABOUT SWALLOWING AN APPLE SEED?
ISBN 0-06-024523-9 (children's hardcover)

FRANNY B. KRANNY, THERE'S A BIRD IN YOUR HAIR!
ISBN 0-06-024683-9 (children's hardcover)

**Visit www.AuthorTracker.com
for exclusive updates on your favorite HarperCollins authors.**

Available wherever books are sold, or call 1-800-331-3761 to order.